Understanding and Enhancing the Impact of Consumer Confidence Reports

The mission of the Awwa Research Foundation (AwwaRF) is to advance the science of water to improve the quality of life. Funded primarily through annual subscription payments from over 1,000 utilities, consulting firms, and manufacturers in North America and abroad, AwwaRF sponsors research on all aspects of drinking water, including supply and resources, treatment, monitoring and analysis, distribution, management, and health effects.

From its headquarters in Denver, Colorado, the AwwaRF staff directs and supports the efforts of over 800 volunteers, who are the heart of the research program. These volunteers, serving on various boards and committees, use their expertise to select and monitor research studies to benefit the entire drinking water community.

Research findings are disseminated through a number of technology transfer activities, including research reports, conferences, videotape summaries, and periodicals.

Understanding and Enhancing the Impact of Consumer Confidence Reports

Prepared by:
Jeffrey K. Lazo, **Joanna L. Pratt**, **Charles N. Herrick**, **Marca L. Hagenstad**, and **Robert S. Raucher**
Stratus Consulting Inc.
P.O. Box 4059, Boulder, CO 80306-4059

Robert E. Hurd
Hurd & Associates, Inc.
107 S. West Street, Alexandria, VA 22314

and

Eric H. Rambo
PA Consulting Group
2711 Allen Blvd, Suite 200, Middleton, WI 53562

Sponsored by:
Awwa Research Foundation
6666 West Quincy Avenue
Denver, CO 80235-3098

Published by:

Awwa Research Foundation
Advancing the Science of Water®

IWA Publishing

DISCLAIMER

This study was funded by the Awwa Research Foundation (AwwaRF). AwwaRF assumes no responsibility for the content of the research study reported in this publication or for the opinions or statements of fact expressed in the report. The mention of trade names for commercial products does not represent or imply the approval or endorsement of AwwaRF. This report is presented solely for informational purposes.

Published by IWA Publishing, Alliance House, 12 Caxton Street, London SW1H 0QS, UK

Telephone: +44 (0) 20 7654 5500; Fax: +44 (0) 20 7654 5555; Email: publications@iwap.co.uk;
Web: **www.iwapublishing.com**

AwwaRF report number 91008F. Originally published by AwwaRF for its subscribers in 2004.
IWA Publishing version published 2005.

ISBN 1 84339 903 2

CONTENTS

TABLES

FOREWORD

The Awwa Research Foundation is a nonprofit corporation that is dedicated to the implementation of a research effort to help utilities respond to regulatory requirements and traditional high-priority concerns of the industry. The research agenda is developed through a process of consultation with subscribers and drinking water professionals. Under the umbrella of a Strategic Research Plan, the Research Advisory Council prioritizes the suggested projects based upon current and future needs, applicability, and past work; the recommendations are forwarded to the Board of Trustees for final selection. The foundation also sponsors research projects through an unsolicited proposal process; the Collaborative Research, Research Applications, and Tailored Collaboration programs; and various joint research efforts with organizations such as the U.S. Environmental Protection Agency, the U.S. Bureau of Reclamation, and the Association of California Water Agencies.

This publication is a result of one of these sponsored studies, and it is hoped that its findings will be applied in communities throughout the world. The following report serves not only as a means of communicating the results of the water industry's centralized research program but also as a tool to enlist the further support of the nonmember utilities and individuals.

Projects are managed closely from their inception to the final report by the foundation's staff and large cadre of volunteers who willingly contribute their time and expertise. The foundation serves a planning and management function and awards contracts to other institutions such as water utilities, universities, and engineering firms. The funding for this research effort comes primarily from the Subscription Program, through which water utilities subscribe to the research program and make an annual payment proportionate to the volume of water they deliver and consultants and manufacturers subscribe based on their annual billings. The program offers a cost-effective and fair method for funding research in the public interest.

A broad spectrum of water supply issues is addressed by the foundation's research agenda: resources, treatment and operations, distribution and storage, water quality and analysis, toxicology, economics, and management. The ultimate purpose of the coordinated effort is to assist water suppliers to provide the highest possible quality of water economically and reliably. The true benefits are realized when the results are implemented at the utility level. The foundation's trustees are pleased to offer this publication as a contribution toward that end.

The reauthorized Safe Drinking Water Act of 1996 added provisions requiring all community water systems (CWS) to deliver annual water quality reports called Consumer Confidence Reports (CCRs) to their customers. With an estimated 46,000 CWS required to develop and distribute CCRs to their customers each year, the CCR requirement poses an important challenge for water utilities to disseminate information to their customers.

The research reported here was conducted to evaluate the effectiveness CCRs and to better understand the impact of CCRs on water utility customers—specifically, to evaluate (1) whether and how CCRs influence consumer perceptions, and (2) what attributes of CCRs most influence consumers' perceptions.

Edmund G. Archuleta, P.E.
Chair, Board of Trustees
Awwa Research Foundation

James F. Manwaring, P.E.
Executive Director
Awwa Research Foundation

ACKNOWLEDGMENTS

This report was funded by the Awwa Research Foundation (AwwaRF). The project was managed and administered by India Williams. The researchers would like to extend their great appreciation to AwwaRF, and especially to India Williams for her management, patience, and encouragement. We also wish to thank the Project Advisory Committee (PAC), which provided very useful comments, guidance, and critiques of this project. The PAC furnished comments on previous reports on this project and generally helped shape and improve this effort. The PAC members were:

Kelly Enders of Medical Education Collaborative
Randolph (Randy) Harmon of Birmingham Water Works
Christine Hoover of The Office of Consumer Advocate, Office of Attorney General, Pennsylvania.

We would also like to thank the participating utilities for their assistance in all phases of this research, their insights into their experiences with Consumer Confidence Reports and utility communications in general, and their thoughts on all aspects of the research, analysis, and reporting of this work:

Laurie Lentz of Austin, Texas
Ed Grusheski of Philadelphia, Pennsylvania
Joe Harbour of Bellevue, Washington
Khis Suravallop of Tacoma, Washington
Sue Atherton of Illinois-American (Peoria, Illinois)
Kathy Allen of Cincinnati, Ohio
Larry Bingaman of Bridgeport, Connecticut—BHC Company; and
Amy Struthers of Boulder, Colorado.

We would also like to thank two external experts who provided input and guidance based on their experiences with water utility communications and reviewed major portions of the report and attempted to gently guide us to correct our errors (any remaining errors are the sole responsibility of the principal investigator):

Branden Johnson, New Jersey Department of Environmental Protection
Elisa Speranza, CH2M HILL

Last, but not least, the researchers wish to thank the many individuals at Stratus Consulting who helped administer the project, execute the technical tasks, and produce this report and Rene Howard of WordProse Inc. for her editorial guidance.

Thank you all very much.

EXECUTIVE SUMMARY

INTRODUCTION

The reauthorized Safe Drinking Water Act of 1996 added provisions requiring all community water systems (CWS) to deliver annual water quality reports called Consumer Confidence Reports (CCRs) to their customers. With an estimated 46,000 CWS required to develop and distribute CCRs to their customers each year, the CCR requirement poses an important challenge for water utilities to disseminate information to their customers. The research reported here was conducted to evaluate the effectiveness of CCRs and to better understand the impact of CCRs on water utility customers—specifically, to evaluate (1) whether and how CCRs influence consumer perceptions, and (2) what attributes of CCRs most influence consumers' perceptions.

METHODS

Following with a review of what is currently known about public perceptions of drinking water safety, the impact of water utility communications on those perceptions, and the research methods used to make those assessments, we used several different techniques to conduct primary research with the aim of better understanding CCR effectiveness. The research comprised five integrated tasks:

1. A mail survey of 118 water utilities exploring what CWS are doing to meet the CCR requirements
2. Ten directed micro-focus groups with water utility customers in five cities to explore customer reactions to reading CCRs
3. CCR attribute characterization, where 127 CCRs were coded on 95 attributes to explore what attributes are most important in determining the usability of CCRs
4. A national random sample of the general population, conducted via a telephone survey with 1,146 water utility customers and 268 well water users to investigate the current reach and impact of CCRs
5. A central site survey with 152 water utility customers in three cites, which evaluated CCRs on multiple response scales to determine how customers react to CCR attributes

RESULTS

Our findings from the research conducted for this project and a review of the existing literature reveal two fundamental "take away" messages:

1. CCRs can and do have a significant and positive impact on customers' perceptions of, attitudes toward, satisfaction with, and trust in their water utility
2. In general, there is room for improving the use of CCRs to communicate more effectively with utility customers

Our findings indicate that CCRs have a net positive impact of 39% on improving customer confidence in the health and safety of their tap water. On several measures of customer satisfaction, trust, and understanding, CCRs can and do have a significant and positive impact on customers' perceptions of their water utilities. However, these impacts depend on the way the CCR is delivered, how it is formatted, and what information it contains.

With about 78% of U.S. households receiving their water from CWS, more than 80 million U.S. households could or should receive a CCR. Our findings indicate that perhaps 40 million people in the United States would have read a CCR from their water utility. With 45% of customers who remembered receiving a CCR indicating that getting the report made them feel more confident about the health and safety of their tap water and 6% saying it made them feel less confident, CCRs have had a net positive impact on about 12 million households nationwide—improving their confidence about the health and safety of their tap water.

Analysis revealed that stylistic attributes of CCRs such as color, photographs, and graphics are significantly related to overall satisfaction. Including informational attributes such as a description of the water treatment process, a frequently asked questions (FAQ) section, informational maps, and a how-to-read-tables section was found to significantly affect satisfaction. The form and content of CCRs, then, do have significant impacts on customers' satisfaction with their water utilities.

RECOMMENDATIONS

We postulated a three-phase model of CCR reading:

initial sort → scanning → reading

Based on this conceptual model of CCR reading, we developed recommendations for water utilities to consider for improving their CCR to make it more effective at each stage of this process.

Initial Sort

With only 38% of customers remembering receiving a CCR, increasing the number of individuals who read their CCR has the greatest potential for having a net positive impact on utility customers. Some approaches to increasing readership include making the CCR look professional and distinct so that it stands out from "junk mail," using a separate mailing rather than a bill stuffer, distributing CCRs to renters and other groups such as schools, nursing homes, and businesses, and expending more effort to reach specific populations such as those living in central cities.

Skimming

Only 45% of customers who remembered getting a CCR said that they read it closely. This suggests that CCRs could be designed to (1) make important information easily accessible for those who will just skim it and (2) include information that would encourage people to read it more carefully. Some approaches to providing this information and increasing the visual appeal of CCRs that we discussed in this research include prominently placing any "take away" messages or statements expressing overall quality and safety, such as "Your tap water is even safer than the state and federal standards for drinking water require," highlighting issues of concern to sensitive subpopulations with any such prominent take away messages, using color, photographs, and maps to draw attention, improve the visual impact, and present information in a way that complements the text.

Reading

When reports are easy to read and understand, they are much more likely to be perceived as being "worthwhile" and to make people feel "confident" that their water is safe and healthy to drink. For individuals willing to make the commitment to reading the CCR more thoroughly, the information should be presented in a credible, accessible, and understandable format. Approaches to improving information communication include having a short but prominent listing of contents, offering an FAQ section or a similarly abbreviated section that allows the customer to quickly locate a particular topic of interest, supplying contact information for individuals who want more information, describing the water treatment process, and keeping information tables as simple as possible, and not avoiding a discussion of contaminants found at measurable levels in the water.

One assumption in undertaking this research has been that utilities are interested in improving customer relations and increasing customer satisfaction with the water utility. While not all of the recommendations outlined above will be applicable in all settings to achieve improved customer satisfaction, the fact that CCRs can and do play a significant role in influencing customer attitudes toward the utility, suggests that many utilities can still benefit from taking a more proactive role in designing, evaluating, and implementing their CCRs. It is hoped that the methods used in this research and the results of our analysis will offer guidance in the quest for more effective CCRs.

CHAPTER 1
INTRODUCTION

This report presents the results of a research project conducted to evaluate the effectiveness of Consumer Confidence Reports (CCRs) and to better understand the impact of CCRs and communications about emerging issues on water utility customers. The research was designed and carried out by Stratus Consulting, Inc., and its research partners (Hurd & Associates, Inc., and PA Consulting, Inc.) for the Awwa Research Foundation (AwwaRF).

This research utilizes a variety of research methods appropriate to the task, including a literature review, customer focus groups, interviews with water utility representatives, an analysis of CCR content and attributes, a national telephone survey, and a quantitative analysis of customer reactions to actual CCRs under experimental conditions.

This study was designed to showcase methods and generate results that local water utilities could apply to make their own CCRs and other communications more effective. This introductory chapter presents background information on CCRs, gives an overview of the research goals and methods, and outlines the report's organization.

RESEARCH GOALS AND OBJECTIVES

In 1996, the U.S. Congress and the President reauthorized and amended the Safe Drinking Water Act (SDWA) by adding a provision requiring that all community water systems (CWS) deliver to their customers an annual water quality report, formally called Consumer Confidence Reports. This provision affected an estimated 46,000 CWS that are now required to develop and distribute CCRs to their customers annually. The first reports were due by October 1999, and the CCRs now come out by July of each year.

The CCR requirement poses both a challenge and an opportunity for water utilities to disseminate valuable and important information to their customers. Although the information in the report is intended to inform the public, it also has the potential to influence their perceptions and attitudes about their local water utility and drinking water. Because some specific types of communication are now mandated, the question for water utilities is no longer *whether* to communicate with their customers, but *how* to do so most effectively.

For that reason, this project is designed to assess the effectiveness and impact of water utility communications, with a specific focus on CCRs as a customer communication tool. Built in part on previous AwwaRF research projects, including a series of consumer attitude surveys, this project qualitatively and quantitatively evaluated (1) whether and how CCRs have influenced consumer perceptions, and (2) what attributes of a utility communication effort have the most influence on consumer perceptions.

Developing effective communications is a deliberate process that can build on past experience as well as the experience of other water utilities and researchers. Toward that end, we sought input from a wide range of relevant sources that included previous research, water utility managers and staff, and the actual customers for whom the communications are intended. Our goals included:

- Reviewing what is currently known about public perceptions of drinking water safety, the impact of water utility communications on those perceptions, and the research methods used to make those assessments
- Determining what water utilities are currently doing to meet the CCR requirements, and what information the CWS would find most helpful
- Reviewing and analyzing the content and attributes of the CCRs that water utilities are currently sending to their customers
- Linking those attributes with customer reactions in an experimental setting to determine what makes a CCR more "usable" for a customer and more "effective" for a water utility
- Assessing the current reach of CCRs and other water utility communications, and the impact these communications are having on customer awareness, perceptions, and attitudes about drinking water safety and their local water utility
- Integrating all of this information into a set of observations, conclusions, and recommendations that local water utilities can apply to evaluate and improve their own communications, including both the mandatory CCR and other voluntary methods

In addition, the project was expanded to help utilities meet the new CCR requirements addressing arsenic and to identify questions and issues that customers and the public are likely to raise about the presence of arsenic in drinking water.

The results of this research give utilities direct information on the effectiveness of different information products and communication approaches. The research also suggests methods and tools that a utility could implement to evaluate the effectiveness of its own communications, especially its CCRs.

RESEARCH METHODS AND PRESENTATION OF RESULTS

To meet these objectives, it was necessary to employ a variety of research methods appropriate to each task, including a literature review, a survey of water utilities, focus groups with customers, an analysis of CCR content and attributes, a national telephone survey, and a "central site survey" that elicited customer reactions to specific types of CCRs.

Our report is divided into eight chapters covering each of the various methods used in this research, and then integrating these results to develop conclusions and recommendations. This first chapter describes the research goals and methods, along with the organization of the report.

Chapter 2 summarizes some of the most relevant research, including studies by the American Water Works Association (AWWA) and AwwaRF on consumer attitudes toward drinking water, on the impact of CCRs and water utility communications on customer perceptions, and on increasing the effectiveness of communicating with customers on water quality issues. This information helped Stratus Consulting and its research partners build on previous research findings and tackle important issues and questions that remain.

Chapter 3 presents the results from a survey of water utility representatives to determine what information would be most useful for improving their communications (particularly with respect to CCRs and emerging issues). The survey of water utilities also elicited useful information about the utilities' experiences in preparing and distributing CCRs and how their customers responded to these efforts.

Chapter 4 describes a series of focus groups conducted by the research team. These discussions were held with water utility customers in five cities across the United States, and were designed to get their reactions to several selected CCRs. This qualitative information was helpful in describing customer perceptions of the various communication materials and in shaping subsequent efforts to quantify customer awareness, perceptions, and reactions.

Chapter 5 presents the results of our efforts to review a broad spectrum of CCRs, and to describe and analyze their contents and attributes. This review and analysis yielded important information about the actual CCRs that customers are receiving, and helped to focus our attention on key attributes that might have a significant impact on customer awareness, perceptions, and reactions in subsequent tasks.

Chapter 6 describes the research methods and results from a national telephone survey to assess public awareness and perceptions of CCRs and other water utility communications, and its impact on other customer attitudes and perceptions. The survey reveals how many customers recall getting a CCR, and how that information affected their perceptions of drinking water safety and their local water utility. Robert Hurd designed the survey to be comparable with other surveys he conducted for AwwaRF in 1993 and 1997-1998, so that we could also track changes in consumer attitudes before and after the CCR requirement was implemented in 1999.

Chapter 7 outlines the research methods and results from the central site survey, in which customers in two cities were asked for their reactions to actual CCRs that contained a mix of attributes identified as potentially important in previous tasks. The project team performed statistical analyses of these customer reactions in an effort to better understand what characteristics help to make a CCR more usable for the customer and more effective for the water utility.

Chapter 8 integrates the key findings and implications from each of the previous methods and chapters, and then presents our conclusions from this study. We also make recommendations to utilities on how they might improve both voluntary and mandated communication programs by applying the results from this study, using these methods to conduct a similar evaluation of their own communications, or both.

Additional information and research materials from this project are appended to this report. Appendix A includes the utility survey reported on in Chapter 3, and Appendix B contains the focus group information and handouts discussed in Chapter 4. Appendix C presents the complete coding scheme for the CCR attribute characterization described in Chapter 5. Appendix D explains the methodology used for the national telephone survey, and Appendix E includes the complete central site survey instrument. Finally, Appendix F contains the complete report on the separate project that examined CCR communication issues and CCR reporting requirements for arsenic (i.e., the “Arsenic Report”).

CHAPTER 2
LITERATURE REVIEW

This chapter reviews and summarizes some previous research of particular relevance to our evaluation of the effectiveness of CCRs and their impact on water utility customers. This review included numerous studies and reports prepared for AWWA and AwwaRF in recent years to measure consumer attitudes about drinking water, to assess the impact of water utility communications, and to identify effective communication practices for drinking water utilities. Our review covers the following topics (listed in order of presentation in this chapter):

- The CCR requirements, water utility efforts, and available resources
- The research methods being used to measure customer awareness and perceptions, and to evaluate the impact and effectiveness of CCRs and other water utility communications
- Qualitative focus groups conducted with water utility customers in developing CCR guidance and preparing for CCR implementation
- Quantitative surveys that measured consumer attitudes, tracked changes over time, and showed the impact of water utility communications
- Other quantitative studies (based on surveys, experiments, or both) that have documented the impact of CCRs specifically, suggested what methods or approaches might be most effective, or both
- Some additional research and reports that offer recommendations and guidance for water utility communications
- A brief discussion of key findings and their implications for this new study to assess the effectiveness of CCRs and their impact on water utility customers

CCR REQUIREMENTS AND WATER UTILITY EFFORTS

Requirements and Resources

As mentioned in Chapter 1, the U.S. Congress and the President reauthorized and amended the SDWA in 1996 by adding a provision requiring that all CWS deliver to their customers an annual water quality report, formally called Consumer Confidence Reports. The CCR rule (40 CFR Parts 141 and 142), issued August 19, 1998, requires CWS to prepare and distribute annual CCRs on the quality of the water delivered by the systems. Every CWS that serves at least 25 residents year-round or that has at least 15 service connections must prepare and distribute a CCR. These systems are diverse, including cities, towns, homeowners' associations, and mobile home parks. Estimates indicate that about 46,000 CWS are required to develop and distribute CCRs to their customers. The first reports were due by October 1999, and they now come out by July of each year.

The rule specifies certain content for the reports. In general, CCRs summarize information that water systems already collect. The report includes basic information on the source(s) of water, the levels of any contaminants detected in the water, and compliance with other drinking water rules, as well as some brief educational material. CCRs aim to help consumers understand the process by which safe drinking water is delivered to their homes and to educate them about the importance of preventive measures—such as source water protection—that ensure a safe drinking water supply. The rationale for CCRs is that consumers have the right to know what is in their drinking water and its source. In the SDWA, the reports are the centerpiece of the public's right to know.

The reports are intended to help customers make informed decisions that may affect their personal health and that of their families and to help them understand the health effects of any contaminants that may be detected in their drinking water. CCRs are also intended to encourage consumers to consider the challenges of delivering safe drinking water, because it is believed that educated consumers are more likely to help protect their drinking water sources and to understand the true costs of safe drinking water. CCRs are among other required communications that utilities must provide to their customers, such as management plans and public notification when MCLs (maximum contaminant levels) are exceeded.[1]

The rule sets baseline standards to ensure that all consumers receive reports that are comparable and include the same type and amount of basic information. However, because the rule does not specify how the reports are to be presented, the approaches to formatting, presenting, and distributing CCRs vary considerably. The amended SDWA also allows states the flexibility to set alternative report requirements after public notice and comment. The basic elements of a CCR must:

- Identify the source of water
- Report contaminants at the highest concentration detected
- Compare the maximum detected concentration with the MCL
- Report violations and relevant information about health effects
- Include boilerplate language with standardized phrases and contact information (Jordan 1998)

The U.S. Environmental Protection Agency (USEPA) has produced guidance for water suppliers, *Preparing Your Drinking Water Consumer Confidence Report*, which is available on the agency's Web site (USEPA 2003a). The guide contains basic information about the reports and answers commonly asked questions such as:

- What content is required in the report?
- What should the report look like?
- How must a water system distribute its report?

The USEPA also developed a Web site called the "CCR iWriter" (USEPA 2003b). This online application enables a utility to produce a regulation-compliant CCR by following instructions and filling in the requested information on each page.

In addition, some states have programs to help utilities understand and comply with the requirement. For instance, the state of Texas sends out standardized templates to utilities to fill out

---

[1] Under the SDWA, to protect the public's health, the levels of specific contaminants in water are limited by setting an enforceable MCL for each regulated contaminant. The MCL is the maximum permissible level of a contaminant in water delivered to any user of a public water system.

(Gary Dixon, Administrative Technician, Texas Natural Resources Conservation Commission, Drink Water Quality Team, personal communication, April 2003).

AWWA has also developed guidebooks (such as AWWA 1998), workshops, and online services to help its water utility members prepare CCRs that comply with federal regulations, but also help to advance water utility goals for effective communications with customers.

Jordan (1998) describes how a water system can produce an effective CCR that is informative and complies with the CCR rule. Jordan offers suggestions on the types of information that should be included, formats that clearly convey information, and methods of distributing the report to all types of consumers.

Implementation by Water Utilities

Several studies, described in the sections that follow, have been conducted in recent years to examine what water utilities are doing to meet the CCR requirements.

Clark et al. (2000) conducted a small survey of water suppliers (n = 50; n = number of responses) across six New England states to assess water utility experience in preparing and distributing the first mandatory CCR in October 1999. Most utilities were able to comply, and many said the experience was not as onerous as they had initially feared. State regulators in the region reported that noncompliance ranged from 25% to 33%, and most of these were small systems serving fewer than 1,000 customers.

Smaller systems said they were either unfamiliar with the requirements or did not have the necessary clerical help or experience, among other factors, for timely compliance. When larger systems were noncompliant, it was usually because of a failure either to follow the mailing and distribution requirements or to submit proper certification to the state. The study also found that compliance increased with the size of the system. Several smaller suppliers expressed frustration that they had to meet the same requirements as the larger suppliers.

Most utility representatives said they recognized the need to reach out and keep customers informed, and supported the CCR in principle. However, many also said they were frustrated by some of the requirements (particularly the language required by the USEPA) and disappointed by the lack of media coverage and customer response.

About half the utilities said they contacted the news media either right before or right after distributing their CCR. About three in four of these utilities were successful in getting some media coverage, although the media often perceived that the results were not particularly "newsworthy." Most often, the media just repeated the information they were given and did no additional reporting. Some suppliers announced the CCR in advance using other methods such as billing inserts, newsletters, separate mailings, and even cable TV. One utility paid for a newspaper advertisement to publicize the CCR results. Clark and colleagues (2000) found no apparent link, however, between advance publicity and either media coverage or customer calls.

About six in ten utilities said they did everything in house. The other four in ten said that they obtained some outside help with writing, design, mailing, or some combination of the tasks involved.

Based on the lessons learned from the survey, the authors made the following suggestions (Clark et al. 2000):

- Simplify technical and statistical language
- Do not hesitate to add to the required language
- Allow plenty of time for contracted services (e.g., writing, design, and mailing), as they can take longer than expected
- Expect media attention for bad news but little recognition of good news
- Accept the frustration of receiving little feedback from customers

Finally, the survey found that some utilities conducted their own customer surveys to identify customer needs and interests either before they prepared their CCR in 1999 or after the CCR was distributed, so they could gauge customer satisfaction with the materials and make improvements for the following year.

The Campaign for Safe and Affordable Drinking Water (CSADW 2000) also conducted an evaluation of the first round of CCRs. The study examined 430 reports in 20 states and the District of Columbia. Although the authors reported that they were impressed with how far utilities had come in producing and distributing reports for the first time, they also identified several areas of concern. They found that 45% of the reports contained unqualified reassurances that the drinking water is "safe," and suggested that this deters people from reading further. In addition, they claimed that the safe statements detract from the purpose of the reports, which is to give consumers a full range of drinking water information so that they can make informed choices and to interest them in efforts to improve water quality.

The USEPA warned in its August 1999 State Implementation Guidance for the reports (USEPA 1999):

> Blanket statements such as "your tap water is safe" may be true for many people drinking the water, but not for members of vulnerable populations such as infants, people undergoing chemotherapy or people with HIV/AIDS. Therefore...be cautious in using the word "safe" and make sure that the required warning statements for vulnerable populations are clearly highlighted in the report.

The CSADW report also found that the required "vulnerable population" warnings were not always prominent—sometimes the warnings were omitted or obscured because of small type or inappropriate placement. The study recommends putting the warning at the beginning of the report, highlighting it with shading or other graphic techniques, and using headings like "Is the Water Safe for Everyone to Drink?" or "Special Information for Immunocompromised People" (CSADW 2000).

OVERVIEW OF RESEARCH METHODS

A number of research methods can be used in various ways to measure customer awareness, perceptions, and attitudes, as well as to evaluate or assess the impact or effectiveness of water utility communications, including the CCR. These methods include:

- Qualitative focus groups
- Quantitative surveys—by phone, by mail, in person, or over the Internet

- Experimental tests
- Other methods for "cognitive mapping" or "mental modeling"[2]
- Less formal methods of getting public and customer feedback through, for example, public meetings

Several studies have conducted focus groups with water utility customers, and assessed public perceptions of drinking water safety in general (Hurd 1993, Owen et al. 1999), or more specifically in terms of perceptions of drinking water safety, the information customers wanted in the CCR, and their preferences for how that information was presented (Hurd and Becker 1998, Macro International, Inc. 1998). Other studies used qualitative focus groups with water utility customers as part of an effort that included quantitative surveys (Hurd 1993, 2003; Colbourne et al. 2001) and helped to illustrate how focus groups can be used in survey development or to complement quantitative methodologies and findings. All of this work shows that focus groups are appropriate and useful methods that can offer important advantages in some specific situations.

A number of studies have used quantitative surveys to measure and assess customer attitudes, perceptions, or the impact of communications, including annual water quality reports (CCRs). Quantitative survey methods are discussed by Hurd (2003) and illustrated by some case studies in Colbourne et al. (2001). Lazo, Kinnel, and Fisher (2000) used a quantitative survey to measure public perceptions of environmental health and safety risks associated with global climate change and other factors. Small et al. (2002) and Lazo, Kinnel, and Fisher (2000) used "mental modeling" to show the differences between expert and lay perceptions and to identify key areas on which to focus for more effective communication.

Several quantitative studies used surveys (either phone or mail) to measure public awareness, perceptions, and attitudes about drinking water quality and safety, and to assess the impact that water utility communications can have on consumer confidence and perceptions. Many of these studies used multiple measures of key independent and dependent variables or multivariate statistical analysis, or both, to identify important relationships and significant differences that help to quantify and explain how these perceptions vary between people, over time, or in response to communications they have received.

In 1993 and again in 1997-1998, AwwaRF sponsored a series of national telephone surveys that was used to measure consumer attitudes and perceptions of drinking water quality, water utility performance, and the information consumers were receiving (Hurd 1993, 1997, 1999, 2003). Hurd used regression analysis to show that perceptions of drinking water quality and safety were among the most important driving factors behind customer satisfaction, and that customers who perceived they were getting the information they needed were significantly more satisfied with their water utility overall (Hurd 1999, 2003).

Johnson (2000) conducted a quantitative mail survey to measure customer attitudes at a New Jersey water utility, and then used an experimental design to show how different changes in

---

[2] Downs and Stea (1973) define cognitive maps as "a process composed of a series of psychological transformations by which an individual acquires, codes, stores, recalls, and decodes information about the relative locations and attributes of phenomena in their everyday spatial environment." Rouse and Morris (1986) define mental models as "the mechanisms whereby humans are able to generate descriptions of system purpose and form, explanation of system functioning and observe system states, and predictions of future system states." For more on cognitive mapping and mental modeling, see Kitchin and Freundschuh 2000, Morgan et al. 2001, and Small et al. 2002.

CCR format and presentation affected customer perceptions of the information and of the safety of their drinking water. Oswald and Hurd (1998) used a quantitative telephone survey and an experimental design to assess the impact of the CCR among customers of a New England water utility. Customers were randomly assigned to two groups, and those interviewed after receiving their CCR were significantly more confident in the safety of their drinking water and significantly more satisfied overall compared to customers who had not yet received a CCR.

As noted earlier, a survey of 50 New England water utilities by Clark et al. (2000) showed that some used surveys to get feedback from their customers to help in developing their CCR or assessing the report's effectiveness after it was distributed. The methodologies and results from these and other studies are explained in more detail in the following sections.

CCR FOCUS GROUPS WITH CUSTOMERS

Both AWWA and the USEPA commissioned focus groups in an effort to anticipate consumer reactions before the final guidance was released and the new CCR requirements were implemented. Other studies, discussed in the sections that follow, have also used focus groups since then to examine customer reactions to specific aspects of the CCRs, including format, language, and other important characteristics.

Hurd and Becker (1998) conducted a series of focus groups for AWWA to better understand the information needs, interests, and preferences of water utility customers, and to gauge their reaction to some actual and prototype CCRs that offered different choices in format, content, and presentation methods. AWWA used the results to provide input from water utilities into the USEPA guidance process, and to recommend how AWWA's member water utilities could use the soon-to-be-required CCRs to communicate more effectively with customers and maximize its potential benefits.

Focus groups were conducted with a cross section of water utility customers in several "test markets," including Southern California, central Wisconsin, and the Philadelphia area. The sites were selected to represent a variety of local demographics, water quality conditions, and levels of customer communications.

Consumer reaction to the CCRs varied significantly from market to market, and appeared to be influenced most strongly by the level of previous utility communication with customers. The study showed that most consumers were interested in getting the information, that some wanted more detail than others, and that many supported utility efforts to get their attention and make the material "stand out."

More than anything else, customers said they wanted reassurance that their water was safe and healthy to drink. Some said that they were not sure the federal standards were sufficient, and that simply meeting those standards would not necessarily make them confident in the water's safety.

Some customers reported that they simply wanted to know that their water was safe, and others wanted to see more supporting detail. The authors said it was clear from customer reactions and comments, however, that "information overload" was a serious concern and that trying to do too much with a single piece of communication could undermine the overall effectiveness and most basic objectives of the CCR.

Most participants said they liked knowing that more information was available and being told where they could get it—if they wanted or needed it.

Most customers felt that they needed independent verification of utility information because of a perceived potential for conflict of interests. This was true even among the most confident and trusting customers. Perceived as most credible were state health departments, area universities, and independent laboratories. News media, the USEPA, and environmental groups were perceived as having their own interests or agendas, making them less credible or reliable sources of information.

Customers generally did not like or understand the label "consumer confidence report," and preferred to call them "water quality reports." They were divided, however, about format and mailing preferences. Some preferred a simple "bill stuffer"; others thought a separate mailing was more likely to get their attention. Some were concerned that bill stuffers would not reach consumers who were renters. Some customers preferred a simple postcard; others wanted a brochure or booklet, or even something they could keep as a reference.

Many said they expected the materials to look professional and that the extra expense was necessary and justified to get their attention—because the information was so important to them. But the authors also said it was possible for materials to be perceived as "too slick," and these customer expectations or standards varied from community to community.

Many liked the use of photos and graphics—as long as they were perceived as "relevant." Some disagreed, but many participants seemed to think that photos of lab personnel and equipment added interest and credibility.

Although most customers were surprised to see their local utility tested for so many different impurities, many found that reassuring. Others said it made them more concerned. There was little agreement as to what contaminants concerned customers most—the chemicals with multisyllabic names or the more familiar substances such as arsenic. Some complained about language they perceived as "alarmist," and others objected to information they found to be "self-serving."

Customers were also divided as to whether they preferred to see the full list of results, or just those where compliance problems were detected. Favorable compromises included adding symbols or highlighting violations on the full list, or presenting separate lists for violations and nonviolations.

Most wanted some narrative to accompany the tables, and wanted larger type to make them easier to read. Many customers preferred the use of whole numbers, and found decimals and exponents confusing—particularly when trying to compare test results against the standards.

The authors observed that water utilities with strong communication and outreach programs seemed to enjoy high levels of customer confidence and support, even in the presence of water quality problems. On the other hand, water utilities that did not appear to have much communication or outreach, were trusted much less by customers—even when they had no previous water quality or compliance problems.

Hurd and Becker (1998) concluded that although the format, style, and content of the CCR was important, the report would be most effective if it were part of a more comprehensive communications strategy and supported by other information, publicity, and outreach activities.

The USEPA commissioned Macro International, Inc. (1998) to conduct a series of focus groups to assess water utility customer information needs, interests, and preferences, as well as to test the readability and appeal of some sample CCR materials. The findings from these focus groups were similar to those from the AWWA focus groups held in that same year (see Hurd and Becker 1998). Three focus groups were conducted with a sample of adult residents from Washington, D.C., that specifically excluded college graduates. A fourth focus group was conducted with eighth graders because the CCR materials were designed for an eighth grade reading level.

Focus group participants were asked a series of questions to determine what they already knew about drinking water, what topics they wanted to know more about, where they look for information, and how they would prefer it be presented. Participants were also shown two different CCRs and asked to react to their appearance, layout, and content, including the language used to describe health effects.

The study found that most customers were interested in, but not very knowledgeable about, their drinking water and its safety. In addition, the study revealed that many were not willing to invest a lot of time and energy to learn about these issues in any detail. Participants indicated that materials should be eye-catching and written in an interesting manner, without scientific jargon.

Participants often said they simply wanted basic information—where their drinking water comes from, whether is "clean" and "safe," and how it is treated. They also wanted information presented clearly and understandably, in "plain English." Many participants said they did not need to know about each and every possible contaminant, but most wanted to know about any substances that violated safety standards.

Most said they would read the water quality brochure, and some said they would be more likely to read the material if it were mailed separately. Some expressed concern that many renters might not receive the information if it was just included in a water bill to the homeowner.

Participants said they preferred to get their information from an independent, "unbiased" third party, and did not necessarily trust either the USEPA or their local water utility to report accurately on safety problems.

Most focus group participants preferred the CCR with photos and tables that were easier to read. They especially liked the section that defined key terms. When presented with alternative phrasings, most preferred shorter, less technical definitions for MCLG (maximum contaminant level goal) and MCL. And many preferred using the word "impurity" rather than "contaminant." After reviewing some sample warnings about possible health effects from contaminants, participants were more concerned when the words "death" or "cancer" were used. They were less concerned when violations were "one-time" problems and when test results were based on animals rather than people (Macro International, Inc. 1998).

Rambo, Baumgartner, and Koenig (2003) recently completed a customer satisfaction study that included some relevant findings about customer communications on water quality issues. A series of focus groups showed that most participants accepted the general idea that water quality information was useful to receive, but relatively few participants were adamant that they wanted more information about the water. There was some sentiment that the water quality information should come from an independent source. In nearly every location, at least one participant expressed skepticism about the accuracy or veracity of water quality information that comes from the utility itself. This appeared to reflect a general mistrust of large organizations. The participants suggested that in areas where people generally trust that their water is safe, they probably would not pay very close attention to more information.

SURVEYS OF CUSTOMER PERCEPTIONS AND COMMUNICATION IMPACT

AwwaRF and AWWA have sponsored several studies that have measured consumer attitudes and perceptions of drinking water issues, tracked changes over time, and identified the key factors that drive consumer attitudes and perceptions. Among these driving factors are water utility communications.

Hurd (1993) conducted a telephone survey with a random sample of 1,600 U.S. adults nationwide as part of a comprehensive study for AwwaRF to measure and analyze public perceptions of and satisfaction with local water utilities, the quality of their drinking water, and the information they receive. The study compared customer expectations and desires against their perceptions of water utility performance, drinking water quality, and information. Statistical analyses, including correlation and regression, were used to understand the interrelationships among various customer perceptions, and to explore the driving factors behind customer satisfaction overall. The results were used to identify specific areas in which utility managers needed to better understand their customers and to consider possible improvements in their products, services, or communications.

As part of this study, a similar and parallel survey was sent to a separate sample that included a cross section of 100 mayors, 50 environmental leaders, and 150 water utility managers. Water utility managers were asked how they thought customers would answer the survey, in an effort to assess the accuracy of the managers' perceptions of their customers. A comparison of the two surveys showed that the water utility managers tended to underestimate public concern about health and safety, and to overestimate the importance of cost or price to the typical consumer. Drinking water quality or safety was the single most important factor consumers said they wanted from their local water utility. About six in ten consumers said they "rarely" or "never" had problems with the taste, odor, or appearance of their tap water; six in ten also rated their water quality as "excellent" or "good."

Even though most people believed that their local tap water complied with federal drinking water standards, about four in ten said that they suspected those standards were "not strict enough." Roughly the same number reported that they had seen, read, or heard something in the media to make them worry that their water might be unsafe to drink.

Most customers gave their local water utility high marks for overall performance. But two thirds of the customers said they received "very little" or "no information at all" from their local water utility, and a majority said it was "very important" that they get more information. Customers also perceived that their local water utilities cared most about "making money" and cared significantly less about drinking water safety and keeping people informed.

Correlation and regression analyses showed that customer perceptions of drinking water quality and safety were important driving factors behind impressions of overall utility performance and satisfaction. Perceptions of water quality were most influenced by the customer's actual experience with the taste, odor, and appearance of tap water. But these perceptions were also influenced significantly by news media reports, and by the extent to which customers perceived the utility as having a positive, "caring" image. When customers perceived that the utility cared about drinking water quality, they were significantly more likely to rate their water quality as good or excellent—even when they experienced problems at the tap. In turn, these perceptions of drinking water quality appeared to be the best predictors of customer satisfaction—and had even more impact than actual problems at the tap or perceptions of price and value.

Hurd (1993) concluded that customer perceptions of drinking water quality and safety were important driving factors behind customer satisfaction, and that utilities had an opportunity to raise customer satisfaction and improve their public image by giving customers more of the drinking water information they said they wanted or needed.

Hurd (1997, 1999, 2003) updated his 1993 consumer attitude survey for AwwaRF by conducting follow-up surveys again in 1997 and 1998. The new surveys were designed to (1) track changes in public attitudes and perceptions since 1993; (2) identify the driving factors behind those changes and customer satisfaction overall; and (3) assess the impact of information consumers were getting from water utilities, the media, and other sources.

The survey also included specific questions about annual water quality reports (CCRs) that were already required in California and distributed voluntarily by some water utilities in other states. In this way, the study was designed to (4) serve as a baseline for measuring the impact and effectiveness of CCRs in future years. In addition, the final report included a handbook with simple answers to common questions on how to conduct survey research for water utilities (Hurd 2003).[3]

Hurd administered the survey by telephone to independent random samples of U.S. adults nationwide in two phases—800 consumers in 1997 and 800 consumers in 1998. When statistical tests showed no significant difference between the two sample years, the results were combined for further analysis into a single sample of 1,600 adults nationwide.

The 1997-1998 survey showed that customer ratings of overall water utility performance had improved significantly since 1993. Follow-up questions and analyses showed other small, but significant, changes that the author identified as contributing factors that together helped explain the increase in overall satisfaction. These factors were (1) improved *perceptions* of water quality, (2) more consumer trust and less skepticism about safety standards, (3) better information and outreach for customers at some utilities, (4) better economic conditions for many customers in the 1997-1998 survey, and (5) improved perceptions of price and value.

Although there was no actual reduction since 1993 in the number of customers who said they experienced problems with taste and odor, there was a small, but statistically significant, improvement in customer perceptions of water quality—affecting about one in ten customers. Small, but statistically significant, shifts of about one in ten customers also emerged on questions measuring consumer trust, the adequacy of federal standards, and local compliance.

Despite the progress in 1997-1998 on some of these key measures, the author concluded that there was more room for improvement. More than three in ten customers said they "sometimes" or "very often" worried that their tap water might not be safe to drink, and many said they wanted more information—as they did in 1993.

About six in ten customers reported that they were receiving little or no information about drinking water quality from their utility or any other source. Follow-up questions showed that three in ten perceived that they were not getting the information they needed to know whether their water was safe to drink.

Although most customers said they perceived no change, about two in ten said their local water utility in 1998 was doing a "better job" of supplying them with information on drinking water compared with five years earlier. In addition, about one in ten customers reported receiving an annual water quality report from their local water utility (either voluntarily or because they lived in California where it had already been required).

Those who reported receiving these water quality reports were more likely than other customers to say: (1) they got "a lot of information" from their utility, (2) they were receiving the information they needed about drinking water safety, (3) their local tap water was "better than"

[3] Preliminary results from phase one showing changes from 1993 to 1997 were presented to AWWA members at their annual conference (Hurd 1997). A subsequent presentation included the phase two results from 1998 and analyzed the impact of water utility communications on customer satisfaction and consumer confidence (Hurd 1999). Complete results and analyses can be found in the final AwwaRF report, which includes the handbook answering basic questions about conducting customer surveys (Hurd 2003).

the federal standards, and (4) their local water utility was doing an excellent job overall (Hurd 1999, 2003).

Further analysis showed that customers who reported getting water quality reports, other information, or both, from their local water utilities were significantly more satisfied than customers who said they did not. Those who felt that they were not receiving the information they needed tended to be less satisfied with their water quality and their local water utility (Hurd 2003).

Hurd also used regression analyses to show the relative impact of the various performance ratings on customer satisfaction overall. The most important driving factors were (1) perceptions of drinking water safety, (2) tap water aesthetics, (3) perceived value and price, and (4) the extent to which the local water utility was perceived as caring about customers. Consumer trust also had a significant impact on satisfaction, even in the presence of other factors. Customer service and public information each had a smaller, but significant, impact as well.

Hurd (2003) concluded that the new CCRs represented an important tool and opportunity for increasing customer satisfaction and improving water utility images while meeting the perceived information needs of customers and complying with new regulatory requirements. Finally, the study suggested that future research use these results as a baseline for tracking changes in customer satisfaction and consumer confidence, and for assessing the impact and effectiveness of CCRs after their nationwide implementation in 1999.

QUANTITATIVE STUDIES OF CCR IMPACT

Several quantitative studies (using surveys, experiments, or both) have been conducted at the local level with the specific purpose of assessing and quantifying the actual impact that CCRs can have on customer awareness, perceptions, and attitudes about drinking water and local water utilities.

Johnson (2000) reports on a mail survey conducted in 1998 with 269 water customers in central New Jersey. This survey was designed to measure customer attitudes, interests, and knowledge about drinking water quality, and to see how differences in CCR format and content affected customer attitudes, perceptions, and reactions. The experimental design systematically varied different features of the CCRs, so that each of the eight survey versions included a different combination of choices for presenting the required CCR information.

The Johnson study found that, at that time, most survey respondents reported that they received little information about drinking water safety and thought it was important to receive more. Statistical analyses showed that customer perceptions of water quality and safety were highly correlated with perceptions of overall utility performance, and that people who received more information were more likely to trust that source of information. Customer ratings across 11 different sources of information showed that the local health department ranked highest; newspapers and TV ranked lowest; and the local water utility was rated somewhere in between. These attitudes and perceptions among New Jersey customers in 1998 were similar to those obtained from national AwwaRF surveys (Hurd 1993, 1997, 1999, 2003).

Most survey respondents were not aware that the utility would soon begin sending water quality reports (CCRs) to all customers, or that such information was already available on request. But most stated that they would read such a report carefully, particularly the first year, and many said that they would visit a utility Web site or call a number listed on the report to get more information.

The survey also measured customer interest in 15 different types of information, 13 of which would be required in the CCR. A factor analysis of these items suggested that they could be divided into two distinct groups—one for narrative information, and one for numerical results about drinking water quality. Regression analyses suggested that demand for numerical results came from customers who were more concerned about drinking water safety and health effects, and who were less trusting of their local water utility.

Johnson found few significant differences in the attitudes, perceptions, or behavior of customers who had received different versions of the hypothetical CCR. With a few minor exceptions, the presentation content and format made no significant difference in whether customers perceived that the information (1) was useful or understandable, (2) made them more concerned or likely to change their behavior, or (3) made them trust the utility any more or less. However, some conditions did make a significant difference in the likelihood that customers would perceive there was a violation, that their perception would be correct, or both:

- Customers receiving a CCR that reported on an actual violation were more likely to say that there was a violation and more likely to be correct. But in either case the numbers were not high. When they had to compare numbers in the table, only 32% correctly saw the violation. Those reading a narrative statement entitled "notice of violation" were only 63% correct.
- Customers receiving CCRs with a nonzero MCLG were more likely to say that there was a violation and more likely to be correct.
- Customers receiving CCRs that included health effects language for a single violation (instead of both items, whether or not there was a violation) were more likely to report that there was a violation and more likely to be correct.
- Customers reading a narrative (instead of a quantitative table) were more likely to say there was a violation, but not any more likely to be correct.
- Customers receiving CCRs with high MCLs (rather than low MCLs) were more likely to report that there was a violation, but significantly less likely to be correct.
- No significant differences were seen between customers receiving CCRs with whole numbers and those reading CCRs with decimals.

Johnson concluded that customers clearly wanted and needed the information that is mandated in CCRs, and that reporting information on contaminants was not likely to have any negative impacts on customer attitudes, perceptions, or behaviors. Some features appeared to make communication more difficult, and even under the best conditions many customers were not able to understand the information correctly. Finally, he questioned whether the less-trusting customers who were most interested in the information would actually believe the report when it reported that the utility had no violations.

Johnson (2001) reports on another survey administered to New Jersey residents in 1999 to more systematically test USEPA-required definitions, language, and information formats for water quality reports. This study found that utility customers had trouble translating the ranges of levels given for detecting contaminants and the health effects language as intended by the agency. Less than half those reading the unexplained information on ranges of contaminant levels correctly said that no violation had occurred. Many people believed that the standard had been violated even if they saw a range with an upper bound below the MCL or a range plus an explanation of why no violation had occurred. As for the health effects language, the "well in excess" phrase evoked larger multiples of the MCL in readers' minds than did the "in excess" language, as the USEPA intended, but nowhere near as much as needed to reflect the actual differences in the agency's calculation of different standards (Johnson 2001).

Oswald and Hurd (1998) conducted a quantitative telephone survey to assess the impact of CCRs and water utility communications on public perceptions of drinking water safety, as well as on customer satisfaction and consumer confidence. This study was based on results from an annual customer survey conducted by a New England water company that had voluntarily begun distributing annual water quality reports (CCRs) 2 years before the reports were required by law. The water company's survey was developed in 1997 to measure customer perceptions, customer satisfaction, and the impact of the utility's communication and outreach efforts. It was administered each year by telephone to a random sample of 800 residential customers, and to a second random sample of commercial customers, as part of an ongoing and comprehensive program to measure and improve customer satisfaction.

Results from the annual customer survey served as the basis for assessing progress toward company quality objectives, identifying opportunities for improvement, and developing strategies and programs to improve internal processes, consumer confidence, and customer satisfaction. The annual survey results were also used to evaluate and improve water utility communication and outreach efforts, including the voluntary CCR.

In 1997, residential customers were divided into two groups and randomly sampled separately. The first group (n = 400) was interviewed before the annual water quality report or CCR was mailed. The second group (n = 400) was interviewed a few weeks later, after customers had received the CCR.

The customer survey results showed that the annual water quality report from the water company had a statistically significant impact—not only on consumer awareness and confidence, but also on broader perceptions and satisfaction with the company overall.

Customers who received a CCR were significantly more satisfied with their water quality, their utility's information efforts, and its performance overall as a water utility. The study also showed that perceptions of water quality and safety are among the most important driving factors behind customer satisfaction.

About four in ten customers recalled reading the 1997 water quality report after it had been mailed. Although customers were exposed to information from the utility and other sources before the mailing, the new water quality report gave a small, but statistically significant, boost to customer awareness and satisfaction with the water company, its product, and its information efforts.

First, customers in the after-mailing sample were significantly more likely than the before-mailing sample to say that they had "seen or heard something about water quality" (32% versus 26%) and to have identified the water company as the source of that information (34% versus 23%). They were also significantly more likely to report that they had read a special mailing from the water company to their home (36% versus 25%).

Second, customers in the after-mailing sample were significantly more satisfied with the company's information efforts compared to the before-mailing sample (with 74% versus 66% being "totally" or "mostly satisfied"), and more likely to say that they were getting the information they needed (72% versus 67%). The after-mailing sample also gave the water company significantly higher ratings for "keeping people informed" about drinking water safety—7.5 versus 6.9 on a scale from 0 (very poor) to 10 (excellent).

Third, customers in the after-mailing sample were significantly more satisfied with the quality of their tap water compared with the before-mailing sample (with 83% versus 79% being "totally" or "mostly satisfied"). It is also worth noting that customers in the after-mailing sample were also significantly more likely to say that their tap water met or exceeded their expectations for taste. Differences in the perceived odor and appearance of their tap water were not statistically significant.

Fourth, customers in the after-mailing sample gave the water company higher ratings for overall performance compared with the before-mailing sample. Before the mailing, 55% said that the company was doing a "good job," and 25% rated the company as "excellent." After the mailing, 53% rated the company as "good" and 31% said that it was doing an "excellent" job—a boost of six percentage points.

Oswald and Hurd (1998) concluded that effective, proactive communications can make a significant difference in customer satisfaction with water utilities, and that the annual water quality reports should be viewed as an opportunity and tool for doing so. The water quality reports gave customers information they needed and wanted, and made them more satisfied with the company, its product, and its information efforts.

Finally, the authors concluded that local water utilities can and should measure and document the impact of CCRs, and that existing customer satisfaction surveys may be used effectively toward that end.

Johnson (2002) conducted two experiments in 2000 to examine the level of customer interest in information that compares their local tap water quality with results from other utilities and from bottled water. While conducting two focus groups in 1996, the author first noticed water utility customers making unsolicited requests for information that compared the quality of local tap water with bottled water (Johnson 1997). In a later survey with customers from a different utility, he also found that more than six in ten wanted a comparison of their own utility's water quality to that of other utilities (Johnson 2000). Johnson conducted the two experiments in 2000 with customers from a New Jersey water utility to assess their interest in comparative information, and to determine what impact that information might have on other customer perceptions. Johnson (2002) found that:

- Utility customers desire information that compares the quality of their local tap water quality with that of other water utilities and of bottled water
- The comparison to other utilities' water quality significantly improved relative ratings of respondents' own tap water
- Comparisons had less effect on consumers' behavioral intentions on bottled water use or home treatment of water than on their views or expressed concerns about water quality

Johnson (N.d.) also conducted research to compare various ways of presenting violations of health standards in CCRs. People reading "qualitative" versions (CCRs that describe water quality without presenting numerical data) were less likely to say that the report gave information on the amounts of substances found in the water. "Extended" reports (those that included reading aids and utility performance information) did only slightly better than the "basic" versions (with minimal information) on overall judgments of water quality and utility performance. Qualitative reports performed worse than others, a finding that supports the decision requiring water utilities to report actual contaminant levels.

Many respondents had trouble identifying the presence or absence of substance amounts or violations when reading the reports, even when the reports provided a bottom line summary on the front page. This suggested that many respondents were not processing this information carefully. The results from this research suggest that water quality reports are no different from other risk communications.

Johnson (N.d.) notes that previous theoretical and empirical research has shown little change in public risk attitudes as a result of one-time or infrequent communications, because of the larger roles of personal experience and prior trust or distrust. And his research suggests that water quality reports are no different in that regard.

ADDITIONAL RESEARCH AND REPORTS

In addition to the research summarized above, several recent (or forthcoming) reports from AWWA and AwwaRF offer suggestions and guidance for more effective communications by drinking water utilities.

Small and colleagues (2002) discuss the content and form of risk communications most likely to improve public understanding of risks from *Cryptosporidium*, as reported in another AwwaRF project: "Protocol for *Cryptosporidium* Risk Communication." This project also yielded basic information on the public's level of understanding of the factors that influence the risk of exposure and infection from *Cryptosporidium*.

The report includes a discussion of risk perception and risk communication. This discussion includes a description of the "mental models" methodology, which can be used to identify the set of issues that define a problem (e.g., risk reduction), and the options available for responding to it. "The goal is to provide information that is technically correct, easy to understand, properly engaging, honest about what is known and what is not, and which will encourage people to act in a prudent and responsible manner, consistent with their self-interest and values" (Small et al. 2002).

Research for the project included open-ended interviews to determine people's knowledge of, misconceptions about, and attitudes toward their drinking water. The interviews were used to serve as a bridge between lay beliefs and the knowledge needed to deal with *Cryptosporidium* risks. On the whole, respondents were satisfied with their drinking water quality, but showed some mistrust of utilities and government agencies, as well as skepticism about these entities' ability to communicate information about drinking water risks honestly and effectively.

The report also discusses available options for utility communications and gives an example and a protocol for developing a *Cryptosporidium* Web page that is available for water utilities to use or incorporate in their communications with the public. To do this, the study identified the most readily accessible information on *Cryptosporidium* and evaluated selected brochures and Web sites from prominent health agencies and water utilities in the United States and Canada.

JohnstonWells (2001) prepared a report for AWWA entitled *Communicating with Customers* based on efforts to develop key consumer messages for the association and its members through industry and public opinion research, AWWA member surveys, focus groups, and analysis. The final report offered communications strategies for water utilities. The authors concluded that consumers have much more knowledge of local drinking water than they do of national drinking water issues, and that AWWA can build on the work that member utilities have done at the local level to communicate industry-wide issues. They also found that most consumers perceive drinking water issues as relatively less serious than other issues such as education, air pollution, crime, traffic, and health care. However, there were some exceptions—for example, with residents in an area of prolonged drought and water restrictions.

In communicating with customers, the report recommends that utilities avoid statements that too easily confuse consumer audiences. For example, they found that customers associate the term "drinking water professionals" with water bottlers and distributors rather than with utilities. Also, the authors report that customers want to know most importantly that their water is safe, even if the source water is not. They want to know that risks are being minimized by using the best technology, and that the water is constantly treated and monitored.

The report recommends including additional explanations to required sections and including a short glossary of terms. It also makes suggestions on how to promote the publication of CCRs through the media.

ETC Institute (N.d.) is conducting research for AwwaRF on customer attitudes, behavior, and the impact of communications efforts, and is preparing a guidebook to give water utilities basic guidelines for developing an effective communications program. This guidebook is expected to

- Identify what factors influence residential water utility customer attitudes and behaviors toward water utilities
- Determine if water utilities can use communication as a tool to positively affect the attitudes and behaviors of residential water utility customers
- Identify the types of information that water utility managers should communicate to enhance customer satisfaction, along with the appropriate methods for communicating such information to customers

Based on information available to date, the ETC Institute research is expected to conclude that customers generally desire information about basic services that water utilities provide—drinking water safety, water availability, and customer billing. If water utilities do not meet these basic informational needs, other communication efforts may not be effective. The guidelines will recommend that water utilities should ensure that routine communications, such as water bills and bill inserts, are designed to communicate these basic types of information to customers.

The guidebook will identify factors that affect customer satisfaction, including communication of information about water quality issues, and recommend ways to inform customers about water quality issues. It will discuss the relative importance that different demographic groups place on different types of information, and then identify which types of messages are most likely to positively influence customer satisfaction among individuals in various demographic groups. The guidebook will also identify the best method of communicating these messages to each of these groups. The project should be completed in 2003.

George Washington University (N.d.) is conducting a project for AwwaRF that will result in strategies and tools for water utilities to communicate information to the public about emerging contaminants more effectively and proactively. The research team has collected and evaluated information from other industries (electric power, chemical, and military) to identify proactive communication strategies and lessons that might be applied to water utilities. Team members have also reviewed literature on situation analysis tools and drafted a diagnostic tool to identify circumstances under which communication about emerging contaminants could be detrimental. Other components of this research are (1) the extraction and scoring of "public concern" attributes, (2) mental model interviews, and (3) Web site analyses to support the team's efforts to prepare new risk communication tools for emerging contaminants. The project should be completed in 2004.

KEY FINDINGS AND IMPLICATIONS

Our review revealed a growing body of research on topics pertinent to our research goals and effort. By necessity, the research team was able to review and summarize only some of the most relevant and readily available material.

Many of the individual studies and reports offer a wealth of detailed information on these issues. We can, however, draw some general conclusions across these studies and discuss in broad terms some important implications for our research here.

Water Utility Efforts and CCR Content

We were unable to find great deal of information about water utility activities or actual CCR content. However, two studies by Clark et al. (2000) and CSADW (2000) helped the project team piece together an initial picture:

- The experience was not as onerous as some New England water suppliers expected. Most were able to comply the first year. Smaller systems had a more difficult time meeting the requirements.
- Water suppliers were most frustrated with the complexity and restrictiveness of the required language.
- Although some suppliers made efforts to publicize the CCRs through the media or in other ways, they were often disappointed with the lack of media coverage or customer calls.
- A few suppliers used customer surveys to help refine their CCRs or assess their actual impact.
- CSADW examined 430 CCRs in 20 states (along with Washington, D.C.) and found that most did a good job of preparing and distributing the reports.
- Nearly half of these, however, gave unqualified reassurances that the water was safe to drink. CSADW was critical of this practice, saying it was not technically accurate and could deter people from reading further.

It would be useful to know more about the experiences of water utilities in preparing and distributing their CCRs, and about the different types of content, format and style that consumers are receiving. Among the most important questions are

1. What options do water utilities have for preparing and distributing the CCRs?
2. What choices do they make, and what is the reasoning behind those decisions?
3. How do those decisions show up in the actual content of the CCRs?
4. How does that ultimately affect consumer awareness, perceptions, or confidence?

The research team was able to address the first two questions in our survey of water utilities (Chapter 3), and the third question in our analysis of CCR content and attributes (Chapter 5). Our review also showed that these methods are appropriate and effective for answering these types of research questions. Questions about actual impact were addressed in the national phone survey (Chapter 6) and the central site survey (Chapter 7).

Customer Focus Groups

Our review showed that qualitative focus groups can be an appropriate and effective method for exploring consumer viewpoints, especially when people are being asked to react to specific communication materials or content.

Despite some important methodological limitations,[4] focus groups can capture a wealth of qualitative information not available from other methods. Focus groups often generate important insights that can later be incorporated and tested in quantitative surveys with large, random samples. In this way, focus groups can result in better survey design or help to illustrate the results of quantitative research.

The USEPA and AWWA initially used focus groups to help shape the final CCR requirements and guidance. Much of this work focused on what the CCR would look like and what it would include, and on how to inform people about a complex technical issue and still produce a report that the typical consumer would find easy to read and understand.

Among the key findings and persistent questions were:

- Consumers are interested in getting information about drinking water quality and safety, but there may be limits. And consumers appear to be divided on the amount of information and technical detail they want, need, or can handle. What is the best way for water utilities to balance these diverse needs and desires?
- More than anything else, consumers said they simply wanted to know whether the water was safe to drink—precisely what some utilities would be criticized for saying (see the CSADW and USEPA work described earlier in this chapter).
- Some consumers said they did not need to know about every possible contaminant. And many said they were not willing to invest a great deal of time and energy. But others said they wanted more detail in terms of quantitative test results and explanations. What is the best way to meet these needs and desires for detail, without creating information overload? One possible compromise is to make sure that customers at least know that the information is available and how they can obtain it.
- The level of information also posed a challenge. Consumers said they wanted the information written in an interesting way, in plain English without any scientific jargon. What is the best way to keep the CCR easy to read and understandable, and still be technically accurate and informative?
- Consumers were also divided on delivery and packaging—whether they preferred a postcard, brochure, or booklet, or whether it was better to send the information along with the water bill (because it was cheaper and easier) or as a separate mailing (to make sure it was noticed and received by all consumers, including renters who do not receive a water bill).

---

[4] Focus group results are often difficult to summarize and compare because they are expressed in words, rather than numbers. More important, they cannot be projected to the total target population because the samples are very small and not necessarily random.

- Many said the information was important to them, and wanted it to be eye-catching and stand out so they would be sure to notice and recognize it. Many said it needed to look professional and attractive to get their attention and be credible, and many preferred to see photos and graphics. But sometimes, materials could also be perceived as too slick, and this appeared to vary by community. Consumers were also sensitive to the tone of the CCR; some materials were perceived as alarmist, and others as self-serving.
- Credibility and trust were among the most important issues. Consumers were often skeptical about information coming from either the water utility or other agencies, including the USEPA. They wanted independent verification by trusted third parties, such as independent labs, universities, or the medical community.
- Some consumers suggested that simply meeting the federal standards would not be enough to make them confident if the standards themselves, the testing, or the reporting were suspect in any way. Even more education and information would be required to address this.
- All these results tended to confirm that the CCRs are more than just a technical exercise in conveying scientific information. Consumers are very sensitive to nuances and subtleties that can affect their perceptions of their drinking water and local water utility.
- Although content, format, and style are important, Hurd and Becker (1998) found that consumer reactions to the sample CCRs were driven primarily by the water utility's prior communications and its relationship with local residents, which varied greatly from community to community. Those with strong communications had more trusting consumers, even when water quality problems had been highly publicized. But a lack of communications seemed to generate mistrust in communities with no history of water quality or compliance problems. That research suggested that CCRs need to be supported by other communications, and would be much more effective if they were part of a more comprehensive communications and outreach strategy.

The focus groups, then, raised some key issues that continue to be relevant, and showed how these methods could be used to reexamine these questions today. The results of the focus groups also helped the research team design and focus the quantitative research conducted later in the project.

Again, our survey of water utilities (Chapter 3) was designed to explore how water utilities answered these questions, and our analysis of CCR content and attributes (Chapter 5) shows how these choices are reflected in the actual content of the CCRs. After reexamining these issues from the consumer perspective using the same focus group methods (Chapter 4), we tested the actual impact of these water utility communications and CCR content in the national telephone survey (Chapter 6) and the central site survey (Chapter 7).

Both of these quantitative tasks incorporated and addressed key issues and factors raised in this review (and the subsequent focus groups).

The national telephone survey asked consumers about all the information they received on water quality (from the media, local utilities, and other sources, in addition to the CCR), and measured perceptions of water quality, trust, and satisfaction with the local water utility. Statistical analysis was then used to identify important relationships between key factors—including communications, trust, confidence, and perceptions of water quality and local water utilities.

The central site survey showed respondents specific CCRs and asked them questions to assess their reactions and link specific CCR attributes with their impact on key perceptions, including readability, credibility, trust, and confidence.

National Surveys

The project team reviewed several national surveys conducted by Robert Hurd for AwwaRF in 1993 and in 1997-1998. These surveys measured consumer perceptions of drinking water quality, available information about drinking water, and local water utilities. They serve as a useful baseline for tracking changes over time and assessing the possible impact of CCRs.

Among the key findings and implications:

- The 1993 survey underscored the value of conducting customer surveys by showing some of the inevitable "gaps" between the full range of consumers (including the "silent majority") and the expectations of water utility managers (shaped by the few who actually call or complain).
- Most consumers in 1993 and 1997-1998 gave their local water utilities high marks for water quality and overall performance. But a majority said they received little or no information about water quality, and wanted more or better information.
- Most consumers said their tap water met (or was better than) the federal standards, but about four in ten said that they were not sure those standards were strict enough. Roughly the same number said they had seen, read, or heard something in the media that made them worry that their local tap water might not be safe or healthy to drink.
- *Perceptions* of drinking water quality and safety were shown to be the most important driving factors behind customer satisfaction and perceptions of overall utility performance.
- Other key drivers were perceptions of value and price, and the extent to which the utility was perceived as caring about customers. Consumer trust also had a significant impact on satisfaction, even in the presence of other factors.
- The 1993 study suggested that water utilities could raise customer satisfaction and improve their public image by giving consumers more of the drinking water information they said they wanted or needed.
- The 1997-1998 survey showed an improvement of 10 percentage points in customer perceptions of overall utility performance. Further analysis traced that to small, but significant, changes in several key factors: (1) improved perceptions of water quality, safety standards, and trust; (2) perceived improvements in communications by customers at some utilities; and (3) improved perceptions of price, value, and the economy in general.
- The 1997-1988 survey showed that two in ten customers thought that their water utilities were doing a better job of keeping them informed, and one in ten reported that they were already getting a CCR (either voluntarily or because they lived in California where it was required).
- Finally, statistical tests showed that consumers who perceived they were getting the information they needed were significantly more satisfied as customers overall. Customers who reported receiving a CCR were also significantly more satisfied.

- The 1997-1998 study concluded that the new CCRs represented an important tool and opportunity for increasing customer satisfaction and improving the image of water utilities while meeting the perceived information needs of customers and complying with new regulatory requirements.

These surveys raised important issues to be addressed in this study, and also provided some useful methods for the national phone survey (Chapter 6) and a baseline against which we could show how consumer perceptions may have changed since the CCRs were required and implemented in 1999. Many of these same questions were repeated again in the 2002 survey, along with new questions designed to get more detailed information about consumer awareness and perceptions of the CCRs.

Other Quantitative Studies

The research team also reviewed several quantitative studies that utilized experimental designs to document more conclusively the impact of CCRs at the local level, and to begin identifying some of the specific components or attributes that lead to success.

Oswald and Hurd (1998) showed that customer surveys can be used to evaluate the effectiveness and impact of CCRs and communications at local water utilities. Customers were divided at random into two groups—one group was surveyed before receiving their CCRs and the other after receiving the report. Those surveyed after getting a CCR were significantly more satisfied with their local water utility, with the quality and safety of their drinking water, and with the utility's information and outreach efforts.

Johnson (2000, 2001, 2002, N.d.) conducted a series of experiments that examined more closely how specific variations in CCR content and presentation can affect consumer reactions and perceptions. Among the key findings:

- Some features made communication more difficult, and even under the best conditions, many customers were unable to understand the information correctly.
- Changes in content and format made no significant difference in the extent to which the information was perceived as useful or understandable. Neither did such changes make customers more concerned, more likely to change their behavior, or more or less trusting of the utility.
- However, some conditions did make a significant difference in customers' tendency to perceive violations, for these perceptions to be accurate, or both.
- Narrative statements helped, up to a point. Only three in ten customers reading a quantitative table were able to identify a violation, but only six in ten could do so when given a narrative statement entitled "notice of violation."
- When given CCRs with nonzero MCLGs and with reading health effects language for violations only (versus language for all contaminants), customers were more likely to perceive violations and to be correct.
- High MCLs (versus low MCLs) and narrative text also made customers more likely to perceive violations, but not more correctly. Presenting the information in whole numbers or decimals made no significant difference.
- Less than half the customers could correctly see that there was no violation when the detection levels were expressed as ranges.

- Other results indicate that customers routinely have trouble interpreting the quantitative tables, but perform even worse when given narrative statements only.

These experiments identify important conditions that can affect customer understanding and perceptions, and provide a useful model for quantifying the impact of specific CCR attributes in the central site survey (Chapter 7). The research team was also able to address some of these features in the survey of water utilities (Chapter 3) and in the analysis of CCR content and attributes (Chapter 5).

Using these methods, we sought to learn more about the choices that water utilities are making and how those choices are reflected in the content of their CCRs. Then, by using similar quantitative methods, we were able to link these attributes to specific customer reactions and perceptions and begin to identify approaches that would make the CCRs more usable for the customer and more effective for the utility. At each step in the process, we tried to build on our review of previous research and design research that would extend beyond our current understanding of these issues.

CHAPTER 3
SURVEY OF WATER UTILITIES

OBJECTIVES

Stratus Consulting developed and implemented a survey of drinking water utilities to explore their experiences and procedures with CCRs and to examine potential future concerns. As we proceeded through the literature review and initial survey design phases of the project, we expanded the survey to address AwwaRF's request for information about the utilities' experiences with, and perceptions of, the USEPA's recent regulation requiring that specific arsenic information be included in CCRs (USEPA 2001).

The specific objectives of the drinking water utility survey that are discussed in this chapter were to collect the following information from the sample utilities:

- Characteristics of the sample utilities
- The utilities' experiences with developing and implementing their CCRs and the public response to these efforts
- Key issues and problems that utilities expect to face in the future (i.e., "emerging issues")
- The utilities' experiences with and perceptions of the newly required arsenic language for CCRs

This chapter presents the findings in the first three categories of information. Appendix F includes an assessment of the utilities' experiences with arsenic requirements.

SURVEY DEVELOPMENT AND IMPLEMENTATION

Stratus Consulting developed a survey questionnaire to obtain the necessary information. The survey was mailed to 401 drinking water utilities, and 118 utilities responded. The methods and issues involved in designing the survey, identifying the sample, and implementing the utility survey are described in the sections that follow.

Survey Development

The utility survey was developed through an iterative process of design and review. Stratus Consulting designed an initial survey based on the four specific objectives described in the previous section. The first series of survey questions elicited information about the water utility, such as type and number of accounts served, sources of water, and issues affecting the utility. The second set of questions asked about the utilities' experiences with developing and implementing their CCRs and the responses they have received from the public. The third series of questions elicited information on the utilities' experience with arsenic, focusing specifically on the arsenic language now required by the USEPA. The fourth group of questions asked about emerging issues that are most important to the utilities. Finally, the research team requested two copies of the utility's most recent CCR.

Stratus Consulting asked the utilities participating in this project to complete the survey and comment on the overall instrument as well as the ease or difficulty of each question. After the project team reviewed this feedback and revised the survey accordingly, the Project Advisory Committee and the AwwaRF project manager reviewed the reworked questionnaire. Finally, the survey was thoroughly reviewed and revised based on feedback and discussion within the Stratus Consulting project team. Appendix A contains a copy of the final survey questionnaire.

Sample Selection

Sample selection involved creating a database of CWS,[5] electing an interim sample of water systems from this database, and selecting a final sample of water systems for the mail survey. Each of these steps is described in the sections that follow.

The first step in selecting a sample of drinking water utilities was to create a database of CWS from which to select our sample. We developed a "compiled database" of CWS using information from two USEPA Microsoft Excel pivot tables, which contain much of the water system information available in the Safe Drinking Water Information System (SDWIS[6]; with the exception of utility contact information):

- SDWIS/FED (the federal version of SDWIS) detailed inventory data, FY2001.xls. This data set contains basic system information, such as system name, location, population served, source water type, and owner type.
- SDWIS/FED violations listed in *Efacts*, Jan 2002.xls. This data set gives information on the frequency and type of violations reported since 1993.

The data were merged to form a single database of 57,383 public water systems. Excluding all noncommunity water systems,[7] any system for which the system name field was empty, and any system for which the owner type was a state or federal government or was not specified left 50,728 CWS. The data for these systems were sorted into 20 groups based on size category (very small, small, medium, large, and very large)[8]; source water type (surface water or groundwater); and whether the system has had any MCL violation(s) since 1993. Thirty facilities were randomly selected from each group, with the exception of the group consisting of very large utilities that rely on groundwater and have had a MCL violation. Because there were only 14 utilities in this

---

[5] A CWS is a "public water system that supplies water to the same population year-round" (USEPA 2002).

[6] The SDWIS "contains information about public water systems and their violations of EPA's drinking water regulations," and can be accessed at http://www.epa.gov/enviro/html/sdwis. SDWIS/FED, the federal version of SDWIS, can be accessed at http://www.epa.gov/safewater/sdwisfed/sdwis.htm.

[7] We excluded Non-Transient Non-Community Water Systems (NTNCWS; defined as public water systems that regularly supply water to at least 25 of the same people at least 6 months per year, but not year-round, including, for example, schools, factories, office buildings, and hospitals that have their own water systems) and Transient Non-Community Water Systems (TNCWS; defined as public water systems that provide water in locations, such as gas stations or campgrounds, where people do not remain for long periods of time).

[8] Very small utilities are defined as those serving 25-500 people. Small utilities serve 501-3,300; medium utilities serve 3,301-10,000; large utilities serve 10,001-100,000; and very large utilities serve more than 100,000 people.

category, all 14 were retained for our sample. This interim sample thus included 584 CWS. Initial contact information was not available from the SDWIS/FED databases used for selecting the interim sample. Consequently, Stratus Consulting filed a Freedom of Information Act (FOIA) request in the spring of 2002 to request contact information for the 584 interim sample facilities. We obtained mailing and telephone contact information for these utilities as requested.

Based on available resources, surveys were mailed to a final sample of 400 CWS. Research team members attempted telephone contacts with the sample of 584 CWS to identify the appropriate contact person at the utility and to confirm the utility's mailing address. Contact information was verified for 228 of the CWS. The final sample included these 228 CWS plus 173 utilities that were randomly selected from the remaining CWS by utility size category. The final sample, then, included 80 utilities from each of the five size categories (plus one utility that was added because of a counting error), for a total of 401 utilities.

Survey Implementation

A survey packet (containing a cover letter, the survey instrument, a flier to remind respondents of key survey requirements, and a prepaid return envelope) was mailed to each of the 401 CWS. Completed surveys were returned by 118 utilities for an overall response rate of 30.8%.[9] Large and very large CWS responded at a higher rate (36% and 49%, respectively) than the smaller utilities (ranging from about 22% to 26%). Table 3.1 shows the number mailed and received, as well as the response rate by size category. The response rate for utilities for which we were able to verify contact information was 34%, compared to a 26% response rate for the utilities for which we had been unable to verify contact information.

A regression analysis (not shown here) of utilities' willingness to return the survey indicated that larger utilities were significantly more likely to return the survey. The utility's use of groundwater or surface water for its source water, its business status (publicly held or privately owned), and its drinking water violation history were insignificant in explaining the utilities' inclination to return the survey. Address verification was also an insignificant factor.

RESULTS AND ANALYSIS

This section summarizes the utility survey results. Information is presented on characteristics of the 118 sample utilities, practices and characteristics of the CCRs developed by the sample utilities, utility perceptions about the value of their CCRs, and past and emerging issues facing the sample utilities. The "n" in each table or row indicates the number of responses evaluated for the statistics presented for each question or item. As not all utilities responded to all questions, the "n" is often less than the total number of utilities in the sample (i.e., less than 118).

---

[9] The U.S. Postal Service returned 18 surveys because they had undeliverable addresses. Consequently, this response rate is based on a total of 383 utilities (i.e., 401 – 18 = 383).

Table 3.1
Surveys mailed and received, and response rate by size category

Size category	Mailed	Total mailed surveys (%)	Responses received	Response rate for size category (%)
Very small	76	19.84	17	22.4
Small	78	20.37	17	21.8
Medium	80	20.89	21	26.3
Large	75	19.58	27	36.0
Very large	74	19.32	36	48.6
Total	383*	100.00	118	30.8

* 401 mailed minus 18 undeliverable addresses resulted in 383 valid addresses.

Table 3.2
Number and types of accounts served

Account type	Mean*	n†	Maximum number of accounts	Total accounts	Percentage of total‡
Households/residential	30,861	95	465,371	2,931,818	88.2
Commercial/industrial	3,727	86	76,000	320,486	9.6
Resale/wholesale to other utilities	13	36	200	474	0.01
Other	3,464	21	25,300	72,739	2.2
Total number of accounts	34,842	118	465,371	4,111,356	100.0

* Summary statistics reported by row do not total by column because many utilities reported only the total number of accounts served and did not identify the accounts by subgroups and also because many utilities reported more than one account type.
† Number of utilities providing information for accounts for each category.
‡ Percentage of totals calculated only for those utilities that reported the number of accounts by subgroups.

Characteristics of Sample Utilities

Size of Utilities (Population Served)

As reported by the utilities, the sample utilities serve about 8.7 million people. The population served per utility ranges from only 64 people to about 1.6 million people, with a mean of more than 121,000 individuals per utility (median of 16,500 people).

Number and Types of Accounts Served

The 118 sample utilities serve a total of 4.1 million accounts. As shown in Table 3.2, the majority of these accounts (88%) are residential customers. About 9.6% of the accounts are commercial accounts and less than 1% (474 accounts) are resale or wholesale to other utilities.

Table 3.3
Type of population area served by sample utilities

	Number of utilities	Percentage of 118 utilities
Small city or town	58	49
Suburban area	40	34
Rural area	38	32
Central city	19	16
Other	4	3
Total*	118	100

*Totals do not add to 118 utilities or 100% because some utilities serve more than one population area.

Two percent of the accounts are classified as "other," including fire protection service accounts, municipal accounts, irrigation and livestock/seasonal accounts, and sprinkler systems.

Source Water. Of the 118 utilities, 115 responded to the question asking them to indicate what percentage of their source water came from groundwater, was derived from surface water, or was purchased. Sixteen utilities (13.9%) reported that they purchase 100% of their water and 83 (72.2%) reported that they purchase none of their water (i.e., they supply 100% of their water). Twenty-three utilities (20% of the total) reported that 100% of their source water was surface water and 48 (41.7%) reported that 100% of their source water was groundwater.

Type of Population Area. The 118 sample utilities serve a variety of different locations, ranging from rural areas to central cities. As Table 3.3 shows, 49% of the utilities serve a small city or town, 34% serve a suburban area, 32% serve a rural area, and 16% serve a central city. Another 3% indicated that they serve other types of locations, including an industrial park, island community, and large county.

Several utilities serve a combination of population areas. Ninety-two of the utilities (78%) reported serving only one type of area. Six of the utilities reported serving all four types of population areas. All these utilities fell into the "very large" size group.

CCR Practices and Characteristics

The utility survey included a series of questions to determine how the sample utilities meet their CCR requirements, including CCR development responsibilities and costs, methods of dissemination, types of publicity, utility additions to the required components of CCRs, and feedback mechanisms. The responses to these questions are described in the next few sections.

CCR Development and Budgeted Costs

The sample utilities were asked to describe who is responsible for designing, preparing, printing, and mailing their CCRs, along with the costs budgeted for these activities. The majority of utilities responded that they design and prepare their CCRs in house, using utility staff and resources. Table 3.4 shows that 68% of the sample utilities indicated that their CCRs are prepared entirely in house. Another 15% of the utilities hire an outside firm to help with CCR design and preparation, but do more than half the work themselves.

Table 3.4
CCR design and preparation responsibilities by size of utility

	Small/medium*		Large†		Total	
	Number	Percentage	Number	Percentage	Number	Percentage
Entirely in house	36	65.5	43	69.4	79	67.5
Mostly in house (i.e., more than 50% of the work completed in house)	6	10.9	11	17.7	17	14.5
Mostly by outside firm (i.e., less than 50% of the work completed in house)	2	3.6	5	8.1	7	6.0
Other	11	20.0	3	4.8	14	12.0
Total	55	100	62	100	117	100

*Includes very small, small, and medium-sized utilities, as defined earlier in this chapter.
†Includes large and very large utilities, as defined earlier in this chapter.

Only 18% of the utilities hire an outside firm to do most (i.e., more than half) or all of their CCR design and preparation, or obtain assistance from other sources. Of these utilities, eight hire an outside firm to do most or all of their CCR preparation, the Texas Natural Resource Conservation Commission prepares CCRs for five Texas utilities, the utilities' water providers design and prepare CCRs for five sample utilities, and state health departments prepare CCRs for three of the utilities. A significant difference was not seen between smaller and larger utilities in terms of how much of the CCR design and preparation work they do in house versus using other sources. The higher percentage of smaller utilities indicating "other" in Table 3.4 largely results from a number of these utilities obtaining all their water from another utility, which also provides their CCR.

Fewer utilities do their own CCR printing and mailing. As Table 3.5 shows, only 38% of the utilities (44 utilities) print and mail their CCRs entirely in house, and 19% (22 utilities) hire an outside firm to assist with some (i.e., less than half) of the work. About 37% of the utilities (43 utilities) hire an outside firm to do more than half of the CCR printing and mailing and another 7% hire an outside firm to do all CCR printing and mailing (seven utilities) or have their water provider print and mail their CCRs (three utilities). Larger utilities are much more likely to use outside resources for CCR printing and mailing than smaller utilities.

On average, the utilities budgeted about $11,400 each year for CCR development, with mean preparation and design costs of about $2,700 per year, printing costs of about $3,700 per year, and mailing costs of about $4,200 per year. However, as shown in Table 3.6, there is a very large range in budgeted CCR expenses (from a total cost of $25 to $180,000 per year) and the median budgeted cost is much lower than the mean.

A key reason for these differences is the variation in budgeted CCR expenses by size of utility. To further explore the costs to utilities of producing and disseminating their CCRs, we performed ordinary least squares (OLS) regression on the cost to utility. Variables used to explain

Table 3.5
CCR printing and mailing responsibilities by size of utility

	Small/medium*		Large†		Total	
	Number	Percentage	Number	Percentage	Number	Percentage
Entirely in house	35	63.6	9	14.5	44	37.6
Mostly in house (i.e., more than 50% of the work completed in house)	7	12.7	15	24.2	22	18.8
Mostly by outside firm (i.e., less than 50% of the work completed in house)	10	18.2	33	53.2	43	36.8
Other	3	5.5	5	8.1	8	6.8
Total	55	100.0	62	100.0	117	100.0

* Includes very small, small, and medium-sized utilities.
† Includes large and very large utilities.

Table 3.6
Amount budgeted each year for CCR development* (n = 106)

	Mean ($)	Median ($)	Range ($)	Standard deviation ($)	n
Mailing costs	4,219	1,300	5-26,132	6,621	77
Printing	3,685	1,500	10-25,000	5,181	84
Preparation and design	2,654	1,000	25-24,500	4,412	47†
Other	1,098	500	0-5,500	1,447	22
Total‡	11,391	3,700	25-180,000	24,088	106

* Reported statistics are results for nonmissing values.
† Less than half the utilities provided budget estimates for their CCR preparation and design expenses (not including "other" expenses), which may be because these costs are more difficult to quantify, are hidden in general labor costs, or both.
‡ Because some utilities reported total amounts only and not the values for individual expense categories, the sum does not equal the total amounts.

costs included the number of CCRs the utility sends, how many different versions of a CCR they produce each year, and whether they include their CCR in their utility bill (as opposed to mailing it separately). Table 3.7 shows the regression results.

With 100 utilities reporting enough information to be included in the regression, 87% of the variance in costs was explained by the model variables. The intercept is not significant, suggesting that most of the variance in cost is captured by the variables used in the regression. The

Table 3.7
OLS regression on cost of CCR production and dissemination (n = 100)
Adjusted r-squared = 0.87

Variable	Estimate	Error	t-value
Intercept	35.95	1,053.94	0.03
Number of CCR versions prepared yearly	3,002.67	195.68	15.34*
Number of CCRs sent each year	0.27	0.032	8.44*
Number of CCRs sent each year squared (divided by 10,000)	-0.004	0.001	-3.30*
CCR included with utility bill	-8,110.94	1,747.71	-4.64*

* Indicates significance at the 1% level.

parameter estimate of "number of CCR versions prepared yearly" of $3,002.67 indicates that, on average, it takes about $3,000 to prepare a CCR (not including printing and mailing costs, which are captured by the next parameters).

The parameter estimate on "number of CCRs sent each year" of 0.27 indicates that it costs about 27 cents per CCR on average to prepare and design, print, and mail a CCR for the first CCR a utility sends. The negative and highly significant estimate on "number of CCRs sent each year squared (divided by 10,000)" indicates that this unit cost falls as the number of CCRs sent increases; in other words, there are economies of scale in CCR implementation. The average utility in the sample sends 35,633 CCRs yearly. Combining the parameter estimates on "number of CCRs sent each year" and "on "number of CCRs sent each year squared (divided by 10,000)," the average cost per CCR sent for the average utility (sending 35,633 CCRs) is 25.4 cents per CCR. According to this estimate, a utility sending 85,633 CCRs rather than the average of 35,633 would pay an average of 23.5 cents per CCR.

Finally, the significant and negative parameter estimate of -$8,110 on "CCR included with utility bill" indicates that utilities that send the CCR as an insert in the utility bill, rather than by first-class, third-class, or bulk mail, save more than $8,000 a year. For an average number of CCRs sent (35,633), this represents a savings of 18.7 cents per CCR. Given that a CCR that can be inserted in a utility bill can be smaller than the average CCR, this savings also likely captures the lower costs of producing a smaller CCR in addition to reductions in mailing costs.

CCR Dissemination

The majority of sample utilities (about 64%) prepares a single version of their CCR each year and sends this version to everyone on their distribution list. A number of utilities (27 of 117) reported preparing no (0) CCRs. Most of these were utilities that purchase all of their water and thus receive a pre-prepared CCR from their water provider. Another 6% of the utilities develop two different versions of their CCR, 3% of the utilities prepare three different CCRs, and 5% develop four or more different CCRs yearly. Presumably, those utilities that prepare more than one CCR per year serve multiple jurisdictions or distinct service areas or sell water to other utilities and develop the CCR for the utility.

On average, each sample utility distributes 35,633 copies of CCRs each year.[10] As would be expected, this number varies by size of utility, from 0 CCRs up to 344,000. The median number of people served by the utility per CCR sent was 2.95. This did not vary significantly by utility size (Kruskal-Wallis test chi-square statistic 3.59, degree of freedom [df] = 4).[11]

About three quarters of the sample utilities (89 or 75.4%) send CCRs to all their water utility accounts. About one quarter of the utilities (25.4%) send CCRs to all the mailing addresses in their service area, allowing them to reach a greater target audience than their account list would provide. Twenty-six percent of the utilities send CCRs to all landlords and property owners in their service area, a practice intended to ensure that building occupants receive a CCR, not necessarily just those who pay the water utility bill (see discussion in Chapter 6 where the national telephone survey suggests that this is not necessarily effective). Thirty-four percent of the utilities indicated that they also distribute CCRs to other audience groups, including:

- Public buildings and offices (16 utilities)
- Apartment complexes (7)
- Libraries (7)
- Post offices (4)
- Retirement homes (4)
- Schools (3)
- Hospitals (2)
- University dormitories, sororities, and fraternities (2)
- Day care centers, factories, gas stations, office towers, public education venues, stores, supermarkets, and mobile home parks (1 each)

The sample utilities use a variety of methods to distribute their CCRs. As shown in Table 3.8, about 34% of the utilities deliver their CCRs by separate bulk mailing, 29% by separate first-class mailing, and 6% by separate third-class mailing. Another 28% of the utilities include the CCR with their utility bills. Most utilities (76%) indicated that they use only one method for delivering their CCRs (either first-class mail, third-class mail, bulk mail, or included with utility bill). Of those that use two mailing approaches, most indicated that they do a first-class mailing in addition to enclosing the CCR in their utility bill—presumably, this approach is undertaken to reach drinking water consumers other than just those who pay the water bill.

---

[10] Calculated as the responses to "How many *people* in total (of all ages) does your utility serve?" divided by "How many total CCRs do you send out each year?"

[11] The Kruskal-Wallis test is used for nonparametric (distribution-free) independent group comparisons. Specifically, it is a nonparametric test used to compare three or more independent groups of sampled data. Unlike parametric independent group tests, the Kruskal-Wallis test makes no assumptions about the distribution of the data (e.g., normality). It is an alternative to the independent group analysis of variance (ANOVA), when the assumption of normality or equality of variance is not met. The Kruskal-Wallis test uses the ranks of the data rather than their raw values to calculate the statistic. The hypotheses for the comparison of two independent groups are

Ho: The samples come from identical populations

and

Ha: The samples come from different populations,

where the hypothesis makes no assumptions about the distribution of the populations. These hypotheses are also sometimes written as testing the equality of the central tendency of the populations.

Table 3.8
Methods of delivering CCRs to the public (n = 116)

	Number of utilities*	Percentage of 116 utilities*
Separate bulk mailing	40	34
Separate first-class mailing	34	29
Included with utility bill	32	28
Separate third-class mailing	7	6
Other	33	28

*Totals do not add to 116 utilities or 100% because some utilities employ more than one delivery method.

In addition, 28% of the utilities stated that they send CCRs by "other" means, including, for example:

- Hand delivery or door-to-door delivery (seven utilities: two are very small, two are small, one is medium-sized, and three are very large—the very large utilities hand deliver CCRs to "some" customers)
- Newspapers (six utilities: they did not indicate whether CCRs are inserted in the newspaper or if a notification of CCR availability is published in the paper)

One utility in each size category included a comment that they send CCRs by other means, such as:

- Utility Web sites
- Inserts in a summer newsletter or local newsletter
- Displays on city hall bulletin boards and utility office counters

CCR Publicity

Each sample utility was asked what methods (other than actually sending the CCR) it uses to publicize its CCRs. As summarized in Table 3.9, one quarter of the sample utilities do not actively publicize their CCRs. The most common methods of advertising CCRs are press releases to the local news media (32% of all sample utilities), public service announcements or ads in local newspapers (30%), and announcements in utility newsletters or bill stuffers (24%). Other types of publicity include community outreach activities and special events (13%), and public service announcements or ads on local television or radio shows (11%).

More than 40% of the sample utilities described other ways in which they publicize their CCRs, including:

- Announcements or fliers posted at utility Web sites (22 utilities), public offices (19), libraries (12), utility offices (4), city Web sites (4), town or community newsletters (3), schools (1), and recreation centers (1)
- Door-to-door delivery of announcements (1)

Table 3.9
Methods for publicizing CCRs (n = 116)

	Number of utilities*	Percentage of 116 utilities*
Press releases to the local news media	37	32
Public service announcements or ads in local newspapers	35	30
Utility newsletter or bill stuffer	28	24
Community outreach activities and special events	15	13
Public service announcements or ads on local television or radio	13	11
Other	48	41
Do not actively publicize CCR	29	25

* Totals do not add to 116 utilities of 100% because some utilities employ more than one publicity method.

- Special meetings or presentations to the city council (1)
- Hand delivery with "Pure Tap" bottles (this utility bottles its own drinking water and distributes it free for special events)

CCR Enhancements

The sample utilities were asked what CCR activities, if any, they perform that are different from most other utilities or go beyond the minimum requirements for CCRs. Responses can be classified in four general groups:

- Use CCR as a marketing or public relations tool
- Employ methods to increase CCR dissemination and readership
- Include written content that is not required by the USEPA
- Include data that are not required by the USEPA

Some of the sample utilities stated that they use their CCR as a tool for marketing their water and wastewater services. These utilities see the CCR as an opportunity for improving public relations by describing their services, system improvements, public education programs, and other information that would be helpful or interesting to their customers.

Utilities mentioned several ways in which they try to increase CCR readership by publicizing and improving the distribution of their CCRs. Most of these approaches involve the publicity methods already described (e.g., by posting the CCR on Web sites, in libraries, and in other public buildings; by airing public service announcements; or by paying for advertisements).

Utilities frequently include information in their CCRs that the USEPA does not require. Topics or materials that may be offered in CCRs include:

- Discussions of current issues (e.g., water quality, water conservation, source water protection, and contaminants such as methyl tertiary butyl ether [MTBE] and arsenic)
- Frequently asked questions
- Educational articles on issues related to water usage and water quality

Table 3.10
Methods used to obtain feedback about CCRs from customers or the public (n = 115)

Feedback method	Number of utilities*	Percentage of 115 utilities*
Provided a phone number, mailing address, or Web site for customer questions or concerns	93	81
Obtained feedback from public meetings or hearings on the topic	14	12
Obtained feedback from customer advisory council or similar group	8	7
Conducted a survey and included questions on the topic	6	5
Conducted focus groups and asked about the topic	1	1
Other	5	4
None of the above	14	12

* Totals do not add to 115 utilities or 100% because some utilities use more than one feedback method.

- Discussions of recent and planned water system improvements
- Descriptions of board activities
- Comment or reply cards (one utility had a 60% response rate from its card)
- Water conservation tips, Web site addresses for additional tips, or both
- Related Web site addresses for kids
- Explanations of the water treatment process
- Additional nonrequired language on contaminants
- Information about watershed education programs
- Descriptions of plans to meet future challenges
- Bilingual information
- Emergency notification telephone numbers to report water problems
- Identification of available water utility services

Additional data included in CCRs include, for example:

- Separate data tables for different wells in the water system (with a cover letter explaining which wells serve which areas)
- All test results (regardless of whether the contaminant is detectable)

Customer Feedback

Most (88%) of the utilities have asked their customers, the public, or both, for feedback about their CCRs (either before or after CCR publication and dissemination), with only about 12% of the utilities reporting that they have not attempted to obtain customer comments. Table 3.10 shows that furnishing contact information (e.g., phone number, mailing address, or

Table 3.11
Percentage of utilities that received customer feedback during the last year by type of comment/question*

Type of comment or question	No comments or questions	One or more comments or questions	A few comments or questions	Several comments or questions	A lot of comments or questions	n
Like the information being provided	25	75	50	21	5	56
Concerns about water quality as reported in CCR	30	70	53	14	4	57
Requested additional copies of the CCR	33	67	53	9	5	43
Too technical or difficult to understand	55	45	38	4	4	53
Questioning spending money on CCRs	72	28	23	6	0	53
Did not receive a CCR	85	15	15	0	0	52

*Values reflect percentage of total number of responses shown in the final column. The last three columns are a breakdown of the column headed "one or more comments or questions."

Web site URL) in the CCR is the most frequent method of obtaining feedback, used by about 80% of the sample utilities. A small number of utilities are proactive in gaining customer reactions to their CCRs. For example, about 12% of the sample utilities hold public meetings or hearings where they obtain client input on CCRs, and 7% obtain feedback from customer advisory councils or similar groups. Although representing a limited percentage, some of the sample utilities have conducted a survey or focus group to obtain feedback about their CCRs (5% and 1%, respectively). For example, one utility includes questions about CCRs in a customer survey that it conducts about every 3 years. A small number of utilities (4%) indicated that they have elicited feedback by other methods, including soliciting comments through the utility newsletter and inviting customers to attend monthly meetings.

The utilities were asked if they received any comments or questions about their CCRs from their customers or the public in the last year. Only 55% (63 utilities) of the 115 sample utilities responded that they had received feedback. We then asked these 63 utilities to indicate how many comments or questions they received related to six types of information. As shown in Table 3.11, only a small percentage of the 63 utilities (i.e., 5% or less) received many questions related to any single type of information. Typically, the utilities received just a few comments or questions in each information category.

The most frequent type of comment the sample utilities received involved positive feedback about CCRs. Seventy-five percent of the sample utilities reported that they had received at least a few comments in the last year from customers saying they like the information supplied in their CCRs. About 70% of the utilities received comments relating to their customers' concerns about the quality of their drinking water, as reported in the CCR. Other comments received by the

Table 3.12
Utility perceptions about the CCR's impact on improving customer perceptions of their utility and drinking water quality

Utility response	Number of utilities	Percentage of utilities*
No impact at all (1)	17	15
Very little impact (2)	40	34
Some impact (3)	50	43
Considerable impact (4)	5	4
Tremendous impact (5)	0	0
Don't know	4	3
Total	116	100

*Does not add to 100% because of rounding.

utilities included requests for additional copies of the CCR (reported by 67% of utilities), complaints that the language was too technical or difficult to understand (45% of utilities), questions about the amount of funds spent to develop the CCRs (28% of utilities), and complaints that the customer did not receive a CCR (15% of utilities).

Utility Perceptions About CCRs

Utilities were asked to describe the impact that their CCRs have had on improving their customers' perceptions of the utility and the quality of their drinking water. Utility respondents were evenly divided between those who believe that CCRs have some or considerable effect on customer perceptions (47%) and those who believe CCRs have very little or no impact (49%) (Table 3.12). The smaller utilities tended to indicate that there would be significantly less impact than the larger utilities (mean rating of 2.19 for the smallest 20% of the utilities versus 2.70 for the largest 20%; Kruskal-Wallis test, chi-square 3.15, df = 1, Pr > chi-square 0.08).[12]

The sample utilities were also asked to indicate the extent to which they agree with a variety of statements about CCRs, water utility communications, and arsenic issues. Table 3.13 summarizes the number and percentage of utilities that responded to each issue and the mean associated with these responses.

It is interesting to note that even though the sample utilities are divided about whether CCRs have improved their customers' perceptions of drinking water quality, most of the utilities (94 utilities, or 80%) agree somewhat or completely that CCRs are a good way to inform customers about drinking water quality. Most sample utilities (100 utilities, or 86%) also agree that CCRs represent only part of their overall responsibility to keep customers informed about water quality.

[12] "df" is, in essence, the number of different things being tested with the statistical test reported. "Pr" is a probability, with a value ranging from zero to one. If the "Pr" value is small, we conclude that the difference between different groups is unlikely to be a coincidence and thus report it as a "statistically significant difference."

Table 3.13
Utility perceptions of water quality and customer communication issues—response frequencies and mean in descending order of mean response*

	Disagree completely (1)	Disagree somewhat (2)	Neither agree or disagree (3)	Agree somewhat (4)	Agree completely (5)	Mean†	n‡
	Percentage of responses						
CCRs are only a part of our overall responsibility to keep customers informed about water quality	3	2	9	32	54	4.3	116
Customers don't care about details of water quality as long as it is safe	8	9	9	34	44	4.1	117
CCRs are a good way to inform customers about drinking water quality	3	3	13	50	30	4.0	117
The newly required arsenic language is important for informing customers about the risks of arsenic in water	4	12	39	32	12	3.4	99
The wording of the required arsenic language will only scare customers	7	9	38	33	14	3.4	113
Overall, providing CCRs to our residential customers has improved our relationship with them	11	10	41	34	4	3.1	115
CCRs do more to scare customers than to inform them	19	16	25	34	6	2.9	116
We don't have sufficient resources to provide our customers all the information we would like them to have	30	22	27	15	5	2.4	113

*Percentages are calculated as the number of utilities responding to each statement divided by the total number of utilities that responded (excluding those utilities that indicated the statement "doesn't apply to my utility").
†Statistics were calculated based on the values "1" through "5" shown in parentheses in the preceding columns.
‡Number of responses other than "doesn't apply to my utility.

The utilities are less sanguine about the extent to which delivering CCRs to their residential customers has improved their relationships with their customers. Only 38% of the sample utilities agree that CCRs help improve customer relationships; 41% are neutral, and 21% disagree.

Almost 80% of the sample utilities agree with the statement, "Customers don't care about details of water quality as long as it is safe." Nine percent neither agree nor disagree with the statement, and 17% disagree. This finding was corroborated by focus group discussions with drinking water customers. Focus group participants indicated that they would like a statement indicating whether their water is safe to drink prominently displayed on their CCR. They were less interested in reading all the details presented in the CCR (see Chapter 4 for additional information). However, as stated in Chapter 2, the USEPA cautions against using such statements in its State Implementation Guidance for CCRs, because it is believed that the "safe" statements detract from the purpose of the reports. The CCRs are designed to give consumers a full range of drinking water information so that they can make informed choices and to interest them in efforts to improve water quality. As discussed in Chapter 2, some groups, such as the NRDC, recommend that CCRs include prominent statements about the utility's compliance with all regulations. If a safe statement is included, it is important to follow up with prominent warning statements for vulnerable populations. As explained in Chapter 2, it has been shown that statements declaring compliance with regulations does help to bring about acceptance and understanding, but that quantitative data are also needed to support this (Johnson 2000, N.d.).[13] This issue is also explored in the central site survey.

Utilities are fairly evenly divided about whether CCRs are more likely to scare customers than inform them. Table 3.13 shows that 40% of the utilities agree that CCRs do more to scare customers than inform them, compared to 35% of the utilities that disagree with this statement. Another 25% of the utilities neither agree nor disagree with the statement.

The sample utilities are somewhat more likely to believe that the newly required arsenic language will scare their customers. About 47% of the sample utilities (48 utilities) agree that the wording of the required arsenic language will scare customers; only 16% disagree with this statement. Interestingly, a similar percentage of sample utilities believe that the newly required arsenic language is nevertheless important for informing customers about the risks of arsenic in water. Of the utilities, 44% (44 utilities) agree with this statement, 16% disagree, and 39% are neutral. These issues are discussed in greater detail in Appendix F, which includes the complete report on the arsenic health effects language study.

Finally, most of the sample utilities believe they have the resources needed to keep their customers informed about the quality of their drinking water. Only 20% of the utilities agreed with the statement, "We don't have sufficient resources to provide our customers all the information we would like them to have." Another 52% of the utilities disagreed with this statement and 27% were neutral. Although marginally still indicating that they have enough resources (mean of 2.96 on a scale where "1" = disagree completely, "3" = neither disagree nor agree, and "5" = agree completely), the smallest 20% of the utilities were less likely to indicate

[13] Others also recommend caution in the use of the word "safe" in describing drinking water quality. For example, the National Resources Defense Council (NRDC 2003) recommends that water systems "Avoid the use of sweeping and prominent claims of absolute safety for their water."

that they had sufficient resources than the largest 20% of the utilities (mean 2.45; Wilcoxon Two-Sample test statistic = 448.00, Z = -1.34, Pr < Z = 0.09).[14]

It is also interesting to note that although the largest 20% of the utilities indicated that they have sufficient resources (mean 2.45), the utilities making up the second largest 20% of utilities were significantly more likely to indicate that they have sufficient resources to give customers the information that they would like them to have (mean 1.68; Kruskal-Wallis test chi-square = 13.78, df = 4, Pr > chi-square 0.008). We cannot tell from the data whether this means that the second largest group of utilities has more resources per customer than the larger utilities, or whether the second largest group of utilities might have a different attitude in terms of wanting to keep their customers informed. If the second largest group of utilities does have more resources per customer, this could suggest some sort of diminishing returns to scale for the largest utilities in terms of information provision.

Past and Emerging Issues and Problems

Past Issues

The sample utilities were asked to rate a number of issues in terms of the extent to which each issue has affected utility operations during the last 3 years. As shown in Table 3.14, regulatory requirements were cited most frequently as having an impact on operations, with 57% of the utilities responding that regulatory requirements have had a moderate, large, or very large effect on operations. Financial and budget issues also presented a problem to the sample utilities in the last 3 years, with 52% responding that these problems created a moderate, large, or very large impact. About 40% of the utilities indicated that old or depleted infrastructure and water system security have had at least a moderate effect on operations in the last 3 years. Significant water shortages and source water protection issues caused moderate to major problems for 33% and 22% of the sample utilities, respectively. Waterborne disease contamination requiring public notification did not create a problem for most utilities. Of the sample utilities, 91% responded that this issue presented no impact at all during the last 3 years.

The impacts of significant drought or water shortage have varied significantly among the different sizes of utilities (Kruskal-Wallis test chi-square 16.44, df = 4, Pr > chi-square 0.003). Interestingly, the smaller utilities report less impact than the larger utilities. The average impact for the smallest 20% of the utilities was 1.52 on a five-point impact scale (1 = no impact at all, 2 = minor impact, 3 = moderate impact, 4 = large impact, and 5 = very large impact)—halfway between none and minor. The average for the largest 20% of the utilities was 2.58—halfway between minor and moderate.

When asked how much source water protection issues had affected or impacted the operations of their water utility in the past three years, the average impact for the smallest 20% of the utilities was 1.48 on the five-point impact scale, suggesting little if any impact for smaller utilities. The average for the largest 20% of the utilities was 2.36. Although this value is still only minor to moderate, it is significantly greater than for the smaller utilities (Kruskal-Wallis test chi-square = 15.28, df = 4, Pr > chi-square = 0.004).

---

[14] In addition to the size categories by which utilities were identified in SDWIS, we categorized utilities by the size of the population served as they reported in the survey. These were then grouped into five groups from smallest to largest (e.g., smallest 20%, second smallest 20%) for additional data analysis.

Table 3.14
Impact of issues on water utility operations over the last 3 years, in descending order from most to least average impact

	No impact at all* (1)	Minor impact* (2)	Moderate impact* (3)	Large impact* (4)	Very large impact* (5)	Mean†	n
	Percentage of responses						
Regulatory requirements	22	21	36	16	5	2.6	117
Financial or budget issues	18	31	36	8	8	2.6	117
Old or depleted infrastructure	30	30	24	12	3	2.3	115
Water system security	37	24	25	12	3	2.2	117
Significant drought/water shortage	46	21	21	7	5	2.0	117
Source water protection issues	43	35	16	5	1	1.9	114
Waterborne disease contamination requiring public notification	91	9	1	0	0	1.1	116

*Percentages based on number of responses listed in the final column.
†Statistics were calculated based on the values "1" through "5" shown in parentheses in the preceding columns.

The largest utilities indicated a moderate to large impact (mean 3.17) of regulatory requirements on their operations in the last 3 years, significantly more than smaller utilities (chi-square 14.32, df = 4, Pr > chi-square = 0.006). Larger utilities felt an even larger impact from water system security issues (mean 3.21), again much more than smaller utilities (mean 1.30) and again, the values were statistically significant (chi-square = 50.35, df = 4, Pr > chi-square < 0.0001).

The impacts on utilities of old or depleted infrastructure, waterborne disease contamination, and financial and budget issues did not vary significantly by utility size.

In summary, regulatory requirements and financial or budget issues have had the largest impact on water utilities in the last 3 years, with regulatory requirements having more of an impact on larger utilities. Significant drought and water shortages, source water protection issues, and waterborne disease contamination requiring public notification have all had only minor or no impact on utilities. But again, where there has been an impact, larger utilities indicated they have felt a greater impact than smaller utilities.

Emerging Issues

The sample utilities were also asked to rate the extent to which they expect various issues to present major concerns in the future. Utilities were asked, "What issues do you see as being potential major concerns in the future of your water utility?" and responded on a four-point scale (1 = "definitely not a future issue for my water utility," 2 = "possibly a future issue for my water utility," 3 = "definitely a minor future issue for my water utility," and 4 = "definitely a major future issue for my water utility"). Table 3.15 shows the four most frequently cited future

Table 3.15
Potential future concerns for water utilities in descending order from most to least concern with the issue

	Definitely not a future issue for my utility* (1)	Possibly a future issue for my utility* (2)	Definitely a minor future issue for my utility* (3)	Definitely a major future issue for my utility* (4)	Mean†	n
	Percentage of responses					
E. Changing regulatory requirements	5	28	34	33	2.94	113
D. Infrastructure replacement and rehabilitation	12	19	38	31	2.88	113
F. Water system security	8	36	33	23	2.71	113
C. Source water protection	12	40	27	22	2.59	111
B. Water supply shortages	18	45	21	16	2.34	109
A. Controlling waterborne diseases	33	46	14	7	1.96	107
G. Providing water in unconventional or new manners (e.g., through in-house filter or bottled water)	59	31	6	3	1.53	108

*Percentages based on the number of responses shown in the last column.
†Statistics were calculated based on the values "1" through "4" shown in parentheses in the preceding columns.

concerns facing utilities—changing regulatory requirements, infrastructure replacement and rehabilitation, water system security, and source water protection. By comparison, the four most frequently cited concerns over the last 3 years were regulatory requirements, financial and budget issues, aging infrastructure, and water system security.

The need for infrastructure replacement and rehabilitation is emerging as a key concern for the sample utilities. Thirty-nine percent of the utilities indicated that aging infrastructure has been a past problem (e.g., a moderate, large, or very large impact in Table 3.14), but 69% anticipate that this will definitely be an issue in the future. Concern about changing regulatory requirements is also increasing, with 67% of the sample utilities expressing concern that changing regulatory requirements will continue to be an issue in the future, compared to 57% of the utilities that responded that regulatory requirements have affected operations in the last 3 years. The largest percentage of water utilities also rated changing regulatory requirements as "definitely a major future issue for my water utility."

Not surprisingly, with continuing apprehension about terrorism, security concerns are emerging as important future issues. Water system security was ranked as the fourth of seven issues affecting utilities in the past, but is ranked as the third of seven concerns for the future. Fifty-six percent of the sample utilities believe that water system security will definitely emerge as a future concern. In addition, 49% indicated that source water protection will be a concern in the future. By comparison, source water protection was ranked as the sixth of seven issues of concern in the last 3 years, with only 22% of the sample utilities responding that this had a moderate, large, or very large impact on operations.

Fewer utilities see water supply shortages (37%) and the need to control waterborne diseases (21%) as definite future issues. Only 9% of the sample utilities are concerned that delivering water in unconventional ways will be a definite problem in the future.

We also compared responses to the questions on potential future concerns between the smallest 20% and the largest 20% of the respondent utilities. Table 3.16 shows the mean response to each issue by utility size. This serves as a statistical test of whether responses between the smaller and large utilities differ significantly. For all the potential future issues, larger utilities were more likely to indicate that they would face these issues in the future. For the larger utilities, water system security was rated as the most likely future issue, followed closely by changing regulatory requirements and infrastructure replacement and rehabilitation. All three of these were rated as between "definitely a minor" and "definitely a major future issue for my water utility," suggesting that these utilities will definitely face these issues in the coming years.

CONCLUSIONS AND IMPLICATIONS

The sample utilities came from a cross section of utility sizes, localities, and water sources. They ranged in size from serving only 64 people to serving about 1.6 million people. Twenty percent indicated that 100% of their source water was surface water and 42% reported that 100% of their source water was groundwater. Almost 50% serve a small city or town, 34% a suburban area, 32% a rural area, and 16% a central city. The majority of the sample utilities' accounts (more than 88%) are residential; only 8% of the utilities' accounts are commercial clients.

According to survey results, most of the sample utilities (68%) design and prepare their own CCRs entirely in house. Another 15% complete more than one half of their design and preparation in house. Fewer utilities are responsible for printing and mailing their CCRs (57% perform all or most of these tasks in house). On average, the utilities budget about $11,000 to design, prepare, print, and mail their CCRs, but there is a very wide range in the budgeted cost for these activities (i.e., from $25 to $180,000). Costs for CCR preparation are largely explained by the number of versions of CCRs a utility prepares and how many are mailed, with some indication of economies of scale in CCR production and dissemination.

Most of the sample utilities (86%) prepare a single version of their CCR each year and send this version to everyone on their distribution list About three quarters of the utilities send their CCR to all their water quality accounts, but about one quarter of the utilities elect to send their CCRs to all mailing addresses in their service area, thereby reaching a greater audience (including, for example, apartment dwellers who do not pay their own water bill). More than one third of the sample utilities also increase their target audience by distributing their CCRs to public buildings, libraries, retirement homes, hospitals, and other locations.

Table 3.16
Comparison of responses between smallest and largest 20% of respondent utilities (shown in descending order of importance to largest utilities) mean/(standard deviation)/n

Statement	Smallest 20%	Largest 20%	Z statistic
Water system security	2.19 (0.98) 21	3.39 (0.66) 23	4.73*
Changing regulatory requirements	2.52 (0.98) 21	3.30 (0.77) 23	2.92*
Infrastructure replacement and rehabilitation	2.52 (1.12) 21	3.22 (0.74) 23	2.40†
Source water protection	2.05 (1.12) 21	2.91 (0.75) 22	2.95‡
Water supply shortages	2.14 (1.06) 21	2.81 (1.078) 21	2.02†
Controlling waterborne diseases	1.52 (0.51) 21	2.19 (0.98) 21	2.76‡
Providing water in unconventional or new manners (e.g., through in-house filters or bottled water)	1.33 (0.58) 21	1.91 (1.02) 22	2.29†

*Value significant at the 1% confidence level.
†Value significant at the 5% confidence level.
‡Value significant at the 10% confidence level.

Methods used for delivering CCRs are fairly evenly divided between bulk mailings (34% of the sample utilities), first-class mailings (29%), and utility bill inserts (28%). In addition, 28% of the sample utilities described a range of other delivery methods including hand delivery, newspaper articles, utility Web sites, inserts in local newsletters, and displays on city hall bulletin boards and utility office counters.

Although about one quarter of the sample utilities do not publicize their CCRs, the remaining utilities employ a variety of publicity techniques, the most frequent of which include press releases to local news media (32% of the sample utilities), public service announcements (30%), and utility newsletter announcements and bill stuffers (24%).

Some sample utilities indicated that they go beyond the basic CCR requirements by including information and data that are not required by the USEPA and employing various publicity methods to increase the dissemination and readership of their CCRs. These utilities indicated that they tend to view their CCRs as an opportunity to improve public relations and to market all their drinking water and wastewater services.

Most of the sample utilities reported that they have attempted to obtain feedback from their customers or the public about their CCRs, and the most frequent feedback method (used by about 80% of the utilities) is to print contact information in the CCR. Twelve percent reported that they held public meetings or hearings to elicit customer feedback, 7% obtained feedback from a customer advisory council or other similar group, and 6% conducted a survey or focus group.

More than one half of the utilities (55%) responded that they have received some type of feedback from their customers. The sample utilities received two types of comments most frequently—positive feedback about the CCR (reported by 75% of the utilities that received feedback) and concerns about drinking water quality (reported by 70% of the utilities).

When asked about their perceptions of the effects of CCRs on various water quality and customer communication issues, the sample utilities generally agreed that CCRs are a good way to inform customers about drinking water quality, although not the only method that utilities should use to keep their customers informed. Nevertheless, the sample utilities were evenly divided about whether CCRs have actually improved their customers' perceptions of the utility and drinking water quality. Similarly, only 38% of the sample utilities agree that CCRs help to improve relationships with customers.

The utilities were fairly evenly divided about whether CCRs are more likely to scare customers than inform them. Forty percent of the utilities agree that CCRs do more to scare customers than inform them, compared to 35% of the utilities that disagree with this statement. Another 25% of the utilities neither agreed nor disagreed with the statement.

When asked about CCR content, almost 80% of the sample utilities agreed with the statement, "Customers don't care about details of water quality as long as it is safe." Nine percent neither agreed nor disagreed with the statement, and 17% disagreed. These results are consistent with findings in prior research (discussed in Chapter 2), where several studies found that customers often primarily are interested in knowing if their water is safe. However, utility representatives must keep in mind that the CCR is to give consumers a full range of drinking water information so that they can make informed choices and to interest them in efforts to improve water quality, and that the USEPA cautions against blanket statements such as "your tap water is safe," and prefers instead to point out compliance to regulations or attach warning statements for vulnerable populations.

When asked to estimate the impact of various issues on water utility operations in the last 3 years, the most frequently cited concerns were regulatory requirements and financial and budget issues, followed by aging infrastructure and water system security. When asked to identify future concerns, utilities identified infrastructure replacement and rehabilitation and changing regulatory requirements as the most critical emerging issues, followed by water system security and source water protection. These responses varied by utility size. Larger utilities considered water system security to be the most likely issue they would face, and smaller utilities indicated that they were more likely to face issues related to changing regulatory requirements or infrastructure replacement and rehabilitation.

The utility survey indicated that water utilities have a very wide range of experience with the development and implementation of their CCRs—often dependent on the size of the utility and the number of customers, as we would expect. There is no overwhelming indication that utilities consider the effort either a waste of time or an extremely valuable exercise—perspectives on the impact of CCRs falls somewhere in between for most utilities. It is apparent that the CCRs require a significant effort on the part of utilities of all sizes (in terms of effort and cost), and also that most utilities realize that this has an impact on their relationships with their customers.

The results of the utility survey paint a portrait of drinking water utility CCR practices and characteristics, and offer utility insights on the extent to which CCRs achieve their communication objectives. Of the 118 utilities that responded to the survey, 117 returned one or more copies of their current CCRs along with the completed survey.

Several characteristics of these CCRs were evaluated in the focus groups described in Chapter 4. These were undertaken in part to determine what CCR characteristics caused individuals to react. The reactions of individuals to the broader set of CCRs (including most of those gathered during the utility survey) are then evaluated further in the central site survey. Our use of these CCRs in subsequent research, especially the CCR attribute characterization and the central site survey, is discussed in subsequent chapters.

CHAPTER 4
FOCUS GROUPS

In July and August 2002 the research team conducted 11 group interviews, or “focus groups,” with residential utility customers in five cities. The objective of the interviews was to elicit evaluative commentary from customers as they read and reacted to CCRs. This information was a vital precursor of and supplement to the national telephone and central site surveys, because it yields important information for survey development and adds qualitative depth to numerical ratings obtained in the surveys.

PROCEDURES

Before the interviews were conducted, a pretest of the discussion guide was done in Madison, Wis., using six temporary employees of PA Consulting Group as participants. This exercise resulted in several small refinements of the guide—for instance, two questions were found to be redundant—but fundamentally it confirmed that the questions were meaningful and would produce useful information.

The interviews were conducted in Boulder, Colo. (three interviews), Philadelphia, Pa. (two interviews), Bellevue and Tacoma, Wash. (two interviews in each city), and Madison, Wis. (two interviews). The first four cities are participating utilities on this project. The fifth, Madison, was added to result in additional geographic diversity and was a convenient site for research access. We did not conduct focus groups in many areas that may display geographic differences. Additional research in other geographic areas and with different size cities (or towns or rural areas) would be beneficial for addressing potentially different issues specific to these “markets.”

Eight participants were recruited for each interview, with an attendance rate of approximately 80% (an average of 6.5 attendees per interview). The participants were homeowners who received water service from the local utility. An effort was made to obtain a roughly representative mixture of ages and education levels, with approximately half being male and half being female. Participants were paid an incentive of $60 for completing the interview.

Each group was asked to read and comment on the CCR produced by its own utility, and either one or two CCRs from other cities. The original design was for each city to review a report from a single additional city from among the cities where focus groups were being held. After the Philadelphia groups, however, the project team determined that the question guide could be streamlined and a third CCR could be reviewed during the interview. In another refinement made after the Philadelphia groups, the order of review was changed so that the local report was no longer automatically reviewed first. This change resulted from the researchers’ concern about the possibility of order effects in the assessment of CCRs.

In a third change after the Philadelphia sessions, a CCR from Nashville, Tenn., was added to the list of reports for review. This CCR was added because all the CCRs from participating utilities were in some ways similar—all were relatively lengthy, with photographs and other characteristics denoting a level of production expense that is not characteristic of all CCRs. The Nashville CCR is a two-page leaflet with more modest production values. The research team determined that this type of report should be reviewed to provide greater contrast in the evaluations. Table 4.1 shows the pattern of review among all cities.

Table 4.1
Schedule of CCR review

City	First review	Second review	Third review
Boulder (July 17, 2002)			
First session	Boulder	Bellevue	Not applicable (N/A)
Second session	Boulder	Philadelphia	N/A
Third session	Boulder	Madison	N/A
Philadelphia (July 27, 2002)			
First session	Philadelphia	Boulder	N/A
Second session	Philadelphia	Bellevue	N/A
Tacoma (August 6, 2002)			
First session	Philadelphia	Tacoma	N/A
Second session	Madison	Tacoma	Nashville
Bellevue (August 7, 2002)			
First session	Nashville	Bellevue	Madison
Second session	Nashville	Bellevue	N/A
Madison (August 14-15, 2002)			
First session	Philadelphia	Madison	Nashville
Second session	Boulder	Madison	Tacoma

QUESTION GUIDE

The interviews began with questions about local water quality, probing on the topics of taste and odor, color and clarity, and concerns about the healthfulness of the water.[15] Next, participants were asked about the information they receive on water quality, and whether they are satisfied with that information. Participants were asked if they had received a report about the quality of their water. They were asked to write down questions they would like to have answered about their water quality.

Next, the interview participants were shown the first CCR and given about 30 seconds to review it. This brief initial review was intended to simulate the kind of cursory initial reading that a report might receive when it arrives in the mail. Initial reactions to the CCR were recorded, and then participants were asked to read the report in more depth, taking 7 to 10 minutes. Participants were asked for their overall reactions to the report. They were asked if the CCR had answered the water quality questions they had raised earlier. They were then sent on a "scavenger hunt" to locate five key pieces of information. Finally, they were probed for more specific reactions to the report, including which information was most important, what effect the report had on their opinion of the utility, and how the report might be improved.

The second and third CCRs were subjected to essentially the same procedure, although the discussion naturally took on a comparative dimension during these reviews.

---

[15] Appendix B contains the moderator's guide for one set of focus groups.

The final portion of each interview discussed the participants' reaction to the new mandatory arsenic language. First, the language for water that fails to meet the standard above 10 μg/L was read, and reactions recorded. Next, the language for water between 5 and 10 μg/L was read, and reactions recorded. Finally, in some instances alternative language that was developed from participants' reactions in the Boulder groups was read. To test whether arsenic is subject to particularly intense concern, in several groups the arsenic language was read, but with trichlorethylene substituted as the contaminant. Appendix F contains the results of the focus group discussions on health effects language.

FINDINGS

The people interviewed for this project were, for the most part, satisfied with their water quality and with the information they received from their water utility about their water. The latter was true even though the majority of participants did not recall receiving a water quality report. Some people were eager to have as much information as possible, but the majority said they primarily wanted to be informed when there is a problem.

At issue has been whether or not to include a list of all potential contaminants or only those that were detected or for which compliance problems were seen. Although not a major focus of the current research, we gleaned from focus group discussions that most people would prefer to have information only on compliance problems and that a complete list of contaminants may be excessive or even potentially confusing.

When participants were asked what information about their water they would like to have, the two most common responses were variations on "What's in it?" and "Is it safe?" Other questions raised about drinking water quality were:

- How and how often is it tested?
- Where does it come from?
- How is our drinking water processed?
- What affects the taste of our drinking water?
- Is there a change in water quality as it is transported to the home?
- How does local water quality compare to other locations?
- How effective are home filtration systems?

As discussed in Chapter 2, some of this information, but not all, is among information required for inclusion in CCRs.

In general, the reaction to the initial CCR reviewed by each group was positive. Participants were generally pleased with the effort to inform customers and appreciated the information. They acknowledged that the report answered most of the questions they had raised. When each was shown the CCR from his or her own city, most said they had never seen it. Opinion was split on whether they had simply failed to notice the report or whether the utility had never in fact mailed one to them.

Only when a second CCR was introduced did the opportunity for comparison create a higher degree of criticism directed toward the reports. To be sure, this is an artificial situation because customers would rarely be in a position to make comparisons of this type. Also, it must be kept in mind that when asked to assert a preference, people typically will do so, whether the preference is strongly or weakly held, and whether the preference is stable over time or highly volatile. These results, then, must be interpreted with caution. Nevertheless, it appeared to the researchers that in most instances the comparative comments emerged spontaneously and naturally, without the moderator having to coax them. This shows that however unnatural the comparison situation may be, it does provide an important opportunity to evaluate the usability and effectiveness of CCRs.

Findings from the focus groups can be organized roughly by a series of four concerns that a water utility could be expected to have about its CCR:

- Is it likely to be noticed and read?
- Does it convey useful information?
- Does it raise concerns about water quality?
- Does it leave a good impression of the utility?

In the following sections we consider focus group findings that bear on each of these questions.

LIKELIHOOD OF NOTICE

Almost all participants agree that a CCR should be attractive and eye-catching to avoid being thrown away as "junk mail." In the course of reviewing the various CCRs, participants frequently remarked that using color, alternative fonts, photos, line art, graphic elements, and simple schematic diagrams helped to make the documents more attractive and noticeable. Comments along these lines helped us develop the list of CCR attributes described in Chapter 5. Such comments also helped to guide survey development and analysis applied in the central site survey discussed in Chapter 7, and to a lesser extent, for the national telephone survey discussed in Chapter 6.

Based on comments made during the focus groups, we postulate a three-phase model of CCR reading.

Initial Sort

When mailed to a customer's home, the CCR enters into a sort of competition with all the other mail the customer receives on a given day. As the focus group participants frequently reminded us, people do not read or even open all their mail. There is an initial sorting process in which people discard what they believe to be "junk" or otherwise irrelevant mail. No matter how important its content, the CCR must be able to survive this initial screen, gain the customer's attention, and make it past the trash can. "Eye-catching" was a phrase describing CCRs that frequently arose in the focus groups. In other words, the CCR should stand out.

Based on these comments it appears that CCRs should (1) clearly identify what it is and who it is from and (2) leave the impression that it is not simply junk mail. The layout provides an initial impression of the water utility. For example, some CCRs may appear to be slick and promotional, leading customers to think they are getting a "sales job." But what appears "slick" to one person may be seen as interesting and inviting to another; for this reason, there can be no hard and fast rules about the best approach to making the "first cut."

Scanning

Even if a CCR survives the initial sort, customers may not read the document word for word. More likely, the customer will give it a cursory scan, looking for red flags or prominent reassurances that everything is basically okay. The document should contain features that facilitate this type of reading, such as:

- A short but prominent listing of contents very near the beginning or initial point of focus, possibly even on the cover
- A question and answer, frequently asked questions (FAQs), or similarly abbreviated section that allows the customer to quickly locate a particular topic of interest
- Prominent placement of any "take away" messages or statements expressing overall quality and safety, such as "Your tap water is even safer than the state and federal standards for drinking water require."

Reading

If a CCR does catch a customer's interest, it is important that it be well written, flow logically from section to section, and avoid relying too heavily on technical terminology, tabular information, or both. A CCR that is eye-catching might be a disappointment if read from cover to cover.

Despite the initial impact of some "high-style" reports, after spending additional time reading, many people preferred other, less visually appealing reports, citing attributes such as clarity, frankness of expression, logical flow, clearer and more intuitive topic headings, and a less distracting format. People sometimes expressed dismissive comments about particularly "slick-looking" documents, dismissing them as marketing or public relations on the part of the utility, and wondering about their excessive cost to produce.

USEFULNESS

No serious problems were encountered in finding information within the reports. Indeed, the research team eliminated the scavenger hunt from the interview when it became clear that most or all participants could locate the key pieces of information. Participants commented favorably, however, when they found a short, to-the-point summary placed prominently on the CCR, telling them that their water is safe. It also seems important that a CCR contain an early and prominently placed explanation of why it is being provided and disseminated. Without reference to USEPA and SDWA requirements, participants tend to speculate about the utility's reasons for producing the report, often ascribing the motivation to marketing or public relations. For some CCRs reviewed, this information was essentially hidden in a small font on the last page and the participants did not see it.

In the comparison portion of the interviews, where more detailed assessments of the CCRs emerged, some reports were criticized for being overtly promotional. Other reports were considered perfunctory, as if they had been done only to comply with federal regulations. Regular comments were heard about the costs of producing reports, with very slick or elaborate reports receiving criticism for being wasteful.

Participants did not appreciate having to stop reading and search for the definition of terms, scanning back and forth between the main narrative and a glossary. Nor did they appreciate a narrative that cannot be read without repeated reference to tables. Because many people are not comfortable with tabular material, it would appear to be best if the narrative were written to stand alone.

The reactions to reports were not always consistent. For instance, the majority opinion in one group was that the Madison, Wis., CCR was poorly organized and difficult to follow. Another group found the same report "professional" and highly informative, despite a few negative comments about its design.

Participants in most of the sessions seem to appreciate a CCR that invites feedback and interaction with the utility. Participants tend to notice and remark favorably on the presence of contact information, feedback forms, explicit invitations to interact with utility representatives, information on the utility's Web site, non-English language contact options, and instructions.

STIMULUS OF CONCERNS

Before reviewing the CCRs, the people interviewed for this project were, for the most part, satisfied with their water quality and with the information their water utility disseminated about their water. In many instances, reading the CCRs introduced participants to water quality issues they had not previously considered. In some cases, the CCR successfully laid these issues to rest; in others, however, concerns lingered. Where they lingered, participants felt that CCRs lacked adequate reassurances about the issues. Often this was connected with seemingly vague wording about the likelihood of injury from a contaminant, or about the level at which a contaminant becomes dangerous. For example, several CCRs raise the issue of potential *Cryptosporidium* contamination, but then leave it to individual readers to determine what—if any—actions should be taken. Participants in several groups referred to this as "CYA" behavior, similar to warning labels on tobacco products. Although not explicitly articulated by participants, it is our sense that people need some type of *actionable* information.

Reactions to the required arsenic language were mixed. One set of participants was not alarmed by it at all. They tended to be aware that the standard had recently been lowered from 50 μg/L, and felt that the old standard had protected people adequately. Another group, however, did find the language troubling. They were especially concerned about the language's uncertainty in terms of the level of risk. Alternative language that was more direct was read to five of the groups. Most preferred this wording for being more honest, although no less troubling.

For a number of participants, wording for the 5-10 μg/L levels was more troubling than the wording for 10+ μg/L level because it contains a passage that makes explicit the economic calculus involved in setting arsenic standards. This placing of a price on human health was found to be unsettling.

IMPRESSIONS OF THE UTILITY DERIVED FROM THE CCR

The general reaction to the initial CCR reviewed by each group was positive. As the dominant sentiment, participants were pleased with the utility's effort to inform customers and appreciated the information. Further discussion, however, brought out a number of dangers to utility credibility and image, most of which we have already cited.

- Documents that were deemed too fancy or expensive-looking sometimes seemed to be too promotional. There seemed to be a fair amount of tolerance for this sin, as organizations are generally expected to promote themselves. However, the participants gave the impression that beyond some ill-defined limit, self promotion would be a significant fault. Self promotion may be viewed as a problem if it raises questions about the report's expense.
- Documents that were not explicit about the reason for distributing the CCR created an opportunity for readers to speculate about the utility's motives, often leading to unflattering conclusions.
- Some CCRs were perceived to contain too much of what was considered "fluff" instead of hard information about water contaminants. In that case, the utility was accused of only going through the motions of providing information, "because they have to," rather than really trying to inform customers. It is not clear from the interviews what lasting impression about the utility this engenders.
- Documents that raised water quality concerns without, in the minds of the readers, setting them to rest, seemed to be evasive. This appeared to stimulate a degree of antagonism toward the utility, creating an impression that it is acting strategically in its own best interest, rather than in the interests of customers.

All these results suggest some room for improvement in CCR communications. In particular, utilities need to consider how the information and its presentation in the CCR is likely to enhance or detract from the trust and satisfaction customers have with their utility.

The focus groups provided ample information for developing the attribute analysis, which is discussed in the next chapter. The research team culled a series of common phrases that were spoken in reaction to the CCRs, which were tested on a larger set of reports and correlated with satisfaction ratings in the central site survey. By modeling these reactions in relation to objective attributes of the CCRs, we expected to obtain valuable insight into the requirements of effective water quality communications.

CHAPTER 5
CCR MULTIATTRIBUTE CHARACTERIZATION

OBJECTIVES

As discussed earlier, the USEPA's CCR regulations, mandated by the 1996 amendments to the SDWA and promulgated by final rule 40 CFR Parts 9, 141, and 142 (USEPA 2001), specify certain information that CCRs must contain to inform customers of their drinking water quality. The information contained in CCRs can raise consumers' awareness of where their water comes from, help them understand the process by which safe drinking water is delivered to their homes, and educate them about the importance of preventive measures—such as source water protection—that ensure a safe drinking water supply. Because the regulations allow for flexibility in terms of report design and additional content, a wide variety of styles and depth of content is seen across the reports. This chapter characterizes and evaluates the array of attributes that can constitute a CCR. The results of this assessment offer utilities ideas for enhancing the usability and effectiveness of their CCRs.

In characterizing CCRs, particular attention has been paid to attributes that capture the usability and effectiveness of CCRs. To customers, a CCR is *usable* if it presents water quality information that may be of value in a way that is easy to understand, easy to remember, and accurate. From the utility's standpoint, a CCR is *effective* for many of the same reasons. It should be noticed amidst the competing appeals for customers' attention, it should be inviting to read, and it should accurately convey information. In addition, however, it should reflect well on the utility, not raising unwarranted concerns about water quality and not calling into question the purposes or professionalism of the organization. The microfocus groups discussed in the previous chapter made it clear that reports can be *usable* without being *effective*. For example, if a CCR presents an appearance of slickness, seems to be a halfhearted effort to fulfill an obligation, or seems to be trying to conceal bad news, customers may not be willing to absorb the information contained in the report.

Typical determinants of usability include format, length, level of interpretation of numerical and technical information, explanation and contextual interpretation of standards, access to supplemental information, content overview or document mapping, provision of general contextual or comparative information, tone of narrative, and logical progression of text and data. All these attributes contribute to the ease of access to information. The determinants of effectiveness are less well understood. Consequently, we have undertaken this study not only to establish which attributes are most important in determining usability, but also to establish the relationship between usability and effectiveness. The information presented in this chapter represents an important input for establishing this relationship.

ATTRIBUTE SELECTION AND ANALYSIS

The research team identified an initial set of 95 attributes that describe the usability or effectiveness, or both, of CCRs. We then selected a smaller subset of attributes from this list that we believe are key measures of a CCR's usability and effectiveness and coded 127 utility CCRs for each attribute. Frequency distributions were conducted for each key attribute and differences resulting from utility size were examined. A more detailed description of our methodology is presented in the sections that follow.

Identification of CCR Attributes

The first step in completing the multiattribute characterization of CCRs involved developing a list of traits that describe CCRs or affect their usability. An initial delineation of attributes was based on:

- Initial review of ten CCRs, selected randomly, to identify characteristics that vary across the reports
- Expert review of the usability of an additional six CCRs, representing a range of geographic regions within the United States, to identify attributes that are likely to affect the ability of consumers to understand the purpose and contents of the reports
- Identification of attributes that respondents mentioned or appeared to respond to in the microfocus groups
- Evaluation of the usability literature to identify attributes that are known to affect the usability of information products (e.g., Hurd and Becker 1998, Zakaluk and Samuels 1988)
- Evaluation of documents identified during the literature review (Task 1) that give insights into the key attributes of CCRs, including Hurd 1993; Hurd and Becker 1998; Macro International, Inc. 1998; Johnson 2000; and Young, Dahan, and Shaffer 2001

Based on this review, we developed a list of 95 CCR attributes and classified them according to four categories:

1. Utility characteristics (e.g., name, address, size, and type of utility)
2. CCR format (e.g., number of pages, font sizes, and inclusion of maps)
3. General content (e.g., date of report, contact information, treatment and monitoring information, and inclusion of water system map)
4. Specific content (e.g., location of the contact information, how water quality data are presented, and information contained in maps)

We developed an approach to score each attribute, based on, for example, a simple *Yes/No* response or a numerical scale. We then tested the initial attributes list by coding a small sample of CCRs and modified the list as needed. Additional revisions were made based on findings from our microfocus group sessions (see Chapter 4). For example, because participants in some of the microfocus groups expressed the desire for a table of contents to enable them to quickly understand the type of information included in the CCR, we added a "Table of Contents" attribute. We also added a measure of readability and a variable measuring the amount of "white space" in the CCR, as described later in this chapter.

Appendix C lists the final CCR attributes identified for the multiattribute characterization, along with the method used to score each attribute. This classification of CCR attributes can be useful to water utilities as they decide how to develop their own CCRs in ways that ensure that their customers notice their CCRs, read them, and understand their content.

Selection of Key Attributes

To narrow and focus our initial set of 95 attributes, we identified a series of key CCR attributes that are most likely to affect usability and effectiveness. These key attributes were selected based on the microfocus group findings (see Chapter 4), our experience in usability analysis, and our review of the usability and CCR literature. Although we do not expect that CCR readers will be altogether aware of which attributes are influencing their reactions, we do anticipate, for instance, that a stylistically appealing report would attract and maintain readers' attention, help readers to retain the information, and possibly convey a positive image for the utility. A report that contains relatively more information would be seen as being more useful, conveying a message that the utility is making a good-faith effort to communicate with customers and is trustworthy. A report that is written in relatively simple language would likewise be seen as more useful and informative, and possibly as being more trustworthy. Certain kinds of content, especially reports of water quality problems, would stimulate concern, but might interact with other composite attributes such as style and informational content to reduce or increase the amount of concern. A prominent initial statement that water quality is sufficiently meeting standards would also be effective in allaying fears.

These conclusions led us to a series of construct attributes that we expect to bear directly on usability and effectiveness: style, informational content, readability, and concern stimulus. These composite attributes were defined as follows:

- Style
  - Text/layout style (textbook/report, newsletter, or brochure)
  - Use of color
  - Presence of photographs
  - Use of graphics
  - Use of "stylish" water system maps (where the map is more for visual appeal than for presenting information)
  - Font size of headings
  - Amount of white space in report (details of this attribute are discussed later)
- Informational content
  - Water supply source
  - Description of standards, what they mean, and how they are implemented
  - Water treatment process description
  - FAQ section
  - "Informational" water system maps (where the map is more for presenting information than for visual appeal)
  - Number of maps
  - Other supplemental information
  - Explanation of how to read data table
  - Description of detected contaminants and sources
- Readability
  - Score of Gunning Fog Index (see explanation that follows)
- Concern stimulus
  - Prominent statement: is your water safe and does it meet standards?

- Educational or health information on *Cryptosporidium*, radon, copper, lead, arsenic, and trihalomethanes
- For how many other contaminants was information easily found?
- What are other contaminants?
- Arsenic level

Two of these attributes, the amount of white space on the page and readability, require additional explanation. White space is defined as the amount of space between the words, lines, and sections in a document. White space can also be found at the edge of the page (to help the reader focus on the important text), and around headings (to help make them stand out). The existence of white space helps highlight the contrasting elements on the page. In addition, a small amount of white space will cause the page to appear gray, dense, and difficult to digest. Newsletter experts recommend using white space as part of newsletter design and state that "negative visual space is as important as the positive" (Daffron 2003). The amount of white space on the sample CCRs was estimated based on a qualitative assessment of the white space on the first two pages (excluding the cover page) of each CCR. The 127 CCRs were then sorted in order of their relative amounts of white space and were rated as having a "low" or a "high" amount of white space.

Readability is defined as the measure of how comfortably or easily text can be read. Several methods exist to estimate the readability of a given text. One of the easiest and most widely used methods is the Gunning Fog Index (Bogert 1985). This index is based on the average number of sentences on each page and the average number of words per sentence with three or more syllables. It is calculated as follows:

$$[(100/\text{the number of sentences}) + \text{the number of words with 3 or more syllables}] \times 0.4$$

A document with a Gunning Fog Index score of 6 to 10 is considered to be easy reading. The average person reads at a Gunning Fog Index score of 9, and anything above 17 is considered to be difficult for university students. The average score for the 60 coded CCRs is 12.5, which is in a medium to difficult reading range. The minimum score is 8.0 and the maximum is 18.0.

Coding CCRs by Attribute

A sample of 127 CCRs was selected for coding by the attributes described in the last section. These data are used as the basis for the CCR analysis described in this chapter, as well as for the central site survey assessment (see Chapter 7). As shown in Table 5.1, these sample CCRs included all 117 CCRs received from the utilities that completed our utility survey; eight CCRs

Table 5.1
CCRs coded in attribute characterization

Utility survey respondents	117
Participating utilities	8
Madison	1
Nashville	1
Total	127

developed by the utilities participating in the project research; and the CCRs developed by the Nashville, Tenn., and Madison, Wis., drinking water utilities (these last two CCRs were included because they were evaluated during our focus group discussions). One analyst coded all 117 CCRs and entered this information to create the CCR Attribute Characterization Database.[16] A second analyst independently coded a subset of these CCRs. The results were compared to ensure accuracy and consistency and to identify any coding problems. Minor discrepancies were identified and corrections were made as needed.

Analytic Methodology

Frequency distributions were conducted for each key attribute to determine the percentage of CCRs that exhibit each attribute. These percentages were also computed by size of utility to determine which CCR characteristics, if any, vary according to utility size. Kruskal-Wallis significance tests were performed to identify any significant differences in CCR attribute by utility size category.

The hypotheses tested in the Kruskal-Wallis test for the comparison of two independent groups are

H_o: The samples come from identical populations

and

H_a: The samples come from different populations,

where the hypothesis makes no assumptions about the distribution of the populations.

Using the Kruskal-Wallis test in this report, we can conclude that significant differences exist by size at a 10% level of confidence when the Pr > chi-square is less than 0.10, at a 5% level of confidence when the Pr > chi-square is less than 0.05, and at a 1% level of confidence when the Pr > chi-square is less than 0.01. A significant Kruskal-Wallis score merely indicates that there is a difference between the different samples—in this case, the different sizes of utilities—on the attribute being examined. Further analysis of the data is necessary to determine the nature of the difference.

RESULTS

Style Attributes

The 127 CCRs were evaluated based on the seven attributes (shown at right and described previously) selected as key descriptors of CCR style. The first style attribute involves the layout of the CCR. The CCRs were assessed to determine if the information was presented in a textbook (e.g., the information is formatted like a report), newsletter (e.g., in columns), or brochure style. Table 5.2 shows that, overall, 50% of the sample CCRs are presented as informational brochures,

Key style attributes

- Text/layout
- Color
- Photographs
- Stylish maps
- Heading font size
- White space

[16] In addition, we coded 60 of the 117 CCRs for all 95 usability and effectiveness attributes. These data were reviewed for the central site survey and are available in the database for use in any additional analysis.

Table 5.2
Type of CCR layout and utility size percentage (n)

	Utility size					
CCR layout	Very small	Small	Medium	Large	Very large	Total
Textbook	7.1 (9)	6.3 (8)	11.0 (14)	5.5 (7)	1.6 (2)	31.5 (40)
Newsletter	1.6 (2)	3.9 (5)	2.4 (3)	6.3 (8)	3.9 (5)	18.1 (23)
Brochure	4.7 (6)	3.2 (4)	3.2 (4)	15.0 (19)	24.4 (31)	50.4 (64)
Total	13.4 (17)	13.4 (17)	16.5 (21)	26.8 (34)	29.9 (38)	100.0 (127)

32% are formatted in a textbook style, and 18% are formatted in a newsletter style. As shown in the table, large and very large utilities (50 of 72, or 69%) tend to produce their CCRs in a brochure format, whereas very small, small, and medium utilities tend to use a textbook layout.

Table 5.3 displays the number and percentage of CCRs that use color and contain photographs, graphics, and stylish maps. As shown in the table, even though the majority (60%) of the CCRs are printed in color, most do not contain photographs, graphics, or stylish maps. About 42% of the CCRs contain one or more photographs, 29% contain one or more graphics, and only 15% include water system maps for stylistic (as opposed to informational) purposes. A significant difference emerges between the smaller and larger utilities, with the large utilities more likely to use color, photos, and graphics.

The font size used for headings can affect both the "look" and readability of a CCR. The CCRs were coded according to whether the headings were in a small font (measured as equivalent to a Times New Roman font of less than 12), medium font (12-14 font size), or large font (≥15 font size). As shown in Table 5.4, just over one half the CCRs used a small font size for headings and 44% used a medium font size. Only 3% of the CCRs used large heading fonts. The CCRs prepared by larger utilities tended to use medium or large font sizes more than the smaller utilities' CCRs (Kruskal-Wallis test, chi-square 20.22, df = 4, Pr > chi-square 0.005).

Like heading font sizes, the amount of white space in a CCR can affect both the look and readability of the CCR. About 43% of the CCRs have a relatively large amount of white space, compared to 57% of CCRs with a small amount of white space. The CCRs prepared by larger utilities tend to use less white space than the CCRs developed by smaller utilities, likely because the larger utilities included more information on average in their CCRs (Kruskal-Wallis test, chi-square 49.55, df = 4, Pr > chi-square < 0.0001).

Based on our analysis of 127 sample CCRs, there is room for improving the appearance of CCRs to make them more attractive and visible to utility customers. Utilities may want to review their CCRs with an eye toward enhancing the overall look of the CCRs through the use of such stylistic attributes as layout, color, photos, graphics, maps, and font sizes.

Table 5.3
CCR style characteristics percentage (n)

Style characteristic	Present in CCR	Not present in CCR	Total	Kruskal-Wallis test for size chi-square (Pr > chi-square) df = 4
Color	60.3	39.4	100	59.31
	(77)	(50)	(127)	< 0.0001*
Photo(s)	41.7	58.3	100	19.17
	(53)	(74)	(127)	0.0007*
Graphic(s)	29.1	70.9	100	24.76
	(37)	(90)	(127)	< 0.0001*
Stylish map(s)	15.0	85.0	100	7.02
	(19)	(108)	(127)	0.13

*Significant at 1% level.

Table 5.4
Font size for CCR headings

Font size	Number of CCRs	Percentage of CCRs
Small	67	52.7
Medium	56	44.1
Large	4	3.2
Total	127	100

Informational Content Attributes

The 127 CCRs were coded based on eight informational content attributes (see Table 5.5). USEPA regulations require CCRs to contain two of these types of information: a description of the utility's water supply source and a description of the contaminants detected in the utility's drinking water (including the source of these contaminants). Table 5.5 shows that about 98% of the CCRs that we evaluated describe the utility's water supply source (the two CCRs that do not contain this information were produced by very small and small utilities). Similarly, close to 95% of the CCRs contain information on the detected contaminants and their sources (the CCRs that do not include this information were produced by utilities in the very small to large size categories). It is interesting to note that the description of detected contaminants is highly visible (i.e., easy to find) in almost all the CCRs examined, because the USEPA requires this information to be placed in a table. By comparison, information on the utility's water supply source is more difficult to locate.

The CCRs also contain a wide variety of information that the USEPA does not require. As depicted in Table 5.5, 28% of the CCRs describe the utility's water treatment process, 21% of the CCRs outline the USEPA water contaminant standards (including what they mean and how they are implemented), 14% explain how to read the data table, 13% of the CCRs present

Table 5.5
CCR informational content percentage (127)

Type of informational content	Not included	Included—somewhat visible	Included—highly visible
Description of water source	1.6 (2)	37.0 (47)	61.4 (78)
Description of detected contaminants and sources	5.5 (7)	0.8 (1)	93.7 (119)
Other supplemental information	44.1 (56)	9.5 (12)	46.5 (59)
Description of water treatment process	71.7 (91)	7.9 (10)	20.5 (26)
Description of standards	78.7 (100)	16.5 (21)	4.7 (6)
Explanation of how to read data table	85.8 (109)	0	14.2 (18)
Informational map	87.4 (111)	0	12.6 (16)
FAQ section	89.0 (113)	0	11.0 (14)

informational map(s), and 11% feature an FAQ section. In addition, about 56% of the CCRs offer other supplemental information (i.e., information that is not included as one of the content attributes listed in Appendix C). The various types of supplemental information contained in more than one CCR are as follows (the number of sample CCRs that provide this information is given in parentheses):

- Conservation tips (18)
- System improvements and future projects (15)
- Source water issues and protection (10)
- Secondary constituents (6)
- Conservation plans and measures (6)
- Water facts, trivia, and history (5)
- CCR certification form (3)
- Sources of additional information; for example, via the Internet (3)
- Water costs and rates (3)
- Water delivery and distribution (3)
- Reliability measures (2)
- Drought concerns and supply shortages (2)

As shown in Table 5.5, with the exception of the description of contaminant standards, the informational content is usually placed in a highly visible location in the CCR so that consumers are likely to see it.

Table 5.6
CCRs with stylish or informational maps, or both

Number of maps	Number of CCRs	Percentage of CCRs
0	92	72.4
1	33	26.0
2	2	1.6
Total	127	100.0

We also examined the number of maps contained in each of the 127 CCRs. Maps ranged from those that are stylish but do not contain detailed water information to maps that are clearly intended to provide such information as the location of wells, surface water sources, pipelines, reservoirs, and treatment facilities. Table 5.6 shows that more than 70% of the CCRs do not contain any type of map. Twenty-six percent of the CCRs (33 CCRs) contain one map and less than 2% (two CCRs, produced by one large and one very large utility) contain two maps. A significant difference emerges between the small and large utilities in terms of their inclusion of maps, with the larger utilities more likely to include one or two maps (Kruskal-Wallis test, chi-square 16.20, df = 4, Pr > chi-square 0.0028).[17]

As described in this section, utilities include a wide variety of supplemental information in their CCRs to disseminate useful information to their customers about their drinking water. The list of supplemental information reviewed in this section and described in Appendix C offers utilities a smorgasbord of ideas for additional CCR content.

Readability

Readability is defined as the measure of how comfortably or easily text can be read. The Gunning Fog Index was used to determine the readability of each CCR's text. This index is based on the average number of sentences on each page and the average number of words per sentence with three or more syllables, as explained previously in this chapter.

A document with a Gunning Fog Index score of 6 to 10 is considered to be easy reading. The average person reads at a Gunning Fog Index score of 9, and anything above 17 is considered to be difficult for university students.

Gunning Fog indices were calculated for all 127 CCRs. The lowest score was 8.1, the highest was 20.5, and the average was 13.1. The scores were categorized into three groups: those below 10 (easy to read), those at 10 and below 17 (medium/difficult), and those at or above 17 (difficult for university students). Seventy-six percent of the CCRs scored in the medium/difficult range, 14% were easy to read, and 10% were difficult (see Table 5.7). Results were also examined by size. Medium-sized facilities produced more easy-to-read CCRs compared to smaller and larger utilities (Kruskal-Wallis test, chi-square 14.62, df = 4, Pr > chi-square 0.006). Otherwise, the results did not seem to indicate that any particular size utility uses easier or more difficult language in their CCRs than the others. Utilities with CCRs that score in the difficult-to-read range may want to consider simplifying their language (e.g., using shorter sentences and words with fewer syllables).

---

[17] Several utilities indicated that they would include less detailed maps as a post 9/11 security measure.

Table 5.7
Readability of CCRs (scores of Gunning Fog Index)

Easy to read (Gunning Fog Index < 10)	14% (18)
Medium difficult to read (10 < Gunning Fog Index < 17)	76% (96)
Difficult to read (Gunning Fog Index > 17)	10% (13)

Concern Stimulus

The concern stimulus attributes are defined as those components of CCRs that could either cause or relieve a customer's concern about drinking water safety. An important concern stimulus attribute is the existence of a statement about regulatory compliance of the utility's drinking water. Focus group participants frequently responded that they would like a statement indicating whether their water is safe to drink prominently displayed in the CCR. Forty-one percent of the CCRs examined contained a highly visible statement that the drinking water is safe or meets all standards, or both (please note that, as discussed in Chapter 2, the USEPA recommends including prominent statements about compliance with standards and advises against statements of safety, as drinking water may not be safe for certain vulnerable populations). Another 39% of the CCRs contained a safety or compliance statement, but the statement was difficult to locate, often hidden in the CCR text. Twenty percent of the CCRs did not include any statement about water safety or compliance with regulations.

The remaining concern stimulus attributes relate to the inclusion of educational or health information on various drinking water contaminants (see Table 5.8). This information can trigger either concern or relief, depending on what the statement says, how it is written, and how the reader interprets it. Nevertheless, the inclusion of health or educational information in a CCR usually suggests that the contaminant is, in fact, present in drinking water, because the USEPA requires this information to be provided when contaminants are detected at specified levels. More CCRs contained information on lead (26% of the 127 CCRs) and *Cryptosporidium* (24%) than any other contaminants. Educational or health information about radon and copper were included in 15% and 13% of the CCRs, respectively. Information about trihalomethanes and arsenic was found in 9% of the CCRs. In addition, 20% of the CCRs contained health or educational information about other contaminants, as follows (the number of CCRs containing information on the contaminant is given in parentheses):

- Nitrate (15)
- Nitrite (4)
- MTBE (3)
- Barium (2)
- Iron (2)
- Sodium (2)

Table 5.8
Percentage of CCRs containing educational or health information on various contaminants percentage %/(n)

Type of contaminant	Educational or health information present	No educational or health information
Lead	26 (33)	74 (94)
Cryptosporidium	24 (31)	76 (96)
Radon	15 (19)	85 (108)
Copper	13 (17)	87 (110)
Trihalomethane	9 (11)	91 (116)
Arsenic	9 (11)	91 (116)
Other contaminants	20 (26)	80 (101)

and one each of:

- 1,1-dichloroethane
- 2,4-D and atrazine
- Alpha emitters
- Beta/photon emitters
- Chromium
- Haloacetic acids
- Hexavalent chromium
- Manganese
- Radium
- Radionuclides
- Zinc

We also examined whether the CCRs specifically listed the arsenic levels (if any) detected in drinking water. Forty-one CCRs (out of 127) listed an arsenic level in their contaminant table. Forty-nine percent of the CCRs showed arsenic levels less than 2 μg/L, 42% had levels greater than 2 and less than 10 μg/L, and 10% had levels exceeding 10 μg/L. See Appendix F for a complete discussion of the new arsenic regulation and its implications for CCRs.

In summary, 81 of the 127 CCRs (64%) contained educational or health information for at least one contaminant. Focus group discussions revealed that concern about drinking water quality is most often ignited by mentioning the presence of contaminants in drinking water, regardless of level. Because information on contaminants often creates, rather than relieves, concern about drinking water safety, we recommend that utilities include a highly visible statement that clearly states whether their water meets the current standards for each contaminant and explains the level of risk posed by each contaminant.

CONCLUSIONS

This chapter has illustrated that, although all CCRs must contain common basic information about water quality, the regulation's flexibility in design has led to a wide array of attributes that can constitute a CCR. These attributes, in turn, affect the usability and effectiveness of CCRs.

One hundred twenty-seven CCRs were examined and evaluated based on four categories of key attributes: style, informational content, readability, and concern stimulus. These key attributes were selected based on the microfocus group findings (see Chapter 4), our experience in usability analysis, and our review of the usability and CCR literature.

Frequency distributions were conducted for each key attribute to determine the percentage of CCRs that exhibit each attribute. These percentages were also computed by size of utility to determine which CCR characteristics, if any, vary according to utility size. Kruskal-Wallis significance tests were performed to identify any significant differences in CCR attribute by utility size category.

Several attributes were examined in the style category. More stylish CCRs are believed to attract the initial interest of more readers, and to maintain that interest, leading the individuals to read further after their initial interest is captured. Fifty percent of the sample CCRs are presented as informational brochures, 32% are formatted in a textbook style, and 18% are formatted in a newsletter style. Although the majority (60%) of the CCRs are printed in color, most do not contain photographs, graphics, or stylish maps. About 42% of the CCRs contain one or more photographs, 29% contain one or more graphics, and only 15% include water system maps for stylistic (as opposed to informational) purposes.

Just over 50% of the CCRs use a small font size for headings and 44% use a medium font size. Only 3% of the CCRs use large heading fonts, which are very helpful for making them stand out and attract readers' attention.

About 57% of the CCRs have a relatively small amount of white space. Although less white space might imply that more in-depth information is included, white space helps to the reader focus on the important text. The lack of white space causes a page to appear dense and difficult to digest.

Our analysis of 127 sample CCRs reveals room for improving the appearance of CCRs to make them more attractive to utility customers. Utilities may want to review their CCRs with an eye toward enhancing the overall look of the CCRs by using such stylistic attributes as enhanced layout styles and larger font sizes for headings, as well as by including color, photos, graphics, and maps.

In the informational content category, utilities were almost all compliant with including the required information of source of water supply and source of contaminants. However, because it was usually placed in a table rather than in the body of text, the information explaining the source of contaminants was easier to find than that covering the source of the water supply.

Most CCRs did not contain information on the water treatment process, the water quality standards, reading of the data table, and answers to FAQs. Just 13% of the CCRs included maps containing information such as the location of wells, surface water sources, pipelines, reservoirs, and treatment facilities. Some CCRs did include additional information such as conservation tips, descriptions of system improvements, and source water issues. The intent of the CCR is to educate consumers on their water quality and build trust in the utility, and we believe that providing more information will help to accomplish these goals. For this reason, we recommend that utilities take advantage of this opportunity to communicate with their customers by including such additional information, even if it goes beyond the regulatory requirements for what a CCR must include.

The readability of the CCRs was evaluated based on the Gunning Fog Index. The average person reads at a Gunning Fog Index score of 9, but the average CCR received a score of 13.1. As revealed in Chapter 3, 45% of the utilities surveyed received complaints that the language used in CCRs was too technical or difficult to understand. Utilities are encouraged to include language aimed at the average reader by keeping sentences short and using words with fewer than three syllables.

Sixty-four percent of the CCRs include educational or health information, or both, for at least one contaminant. Even though including information on contaminants is encouraged and often required (if a contaminant level exceeds the standard), it should be presented to stimulate more reassurance than concern. About 70% of the utilities reported receiving comments on their customers' concerns about the quality of their drinking water as reported in the CCR (see Chapter 3). This is attributable to the discussion of contaminants, even though the utility may be meeting all standards.

The attribute analysis conducted in this part of the research project was used for designing the central site survey, where utility customers examined and rated CCRs, and for analyzing the survey results (see Chapter 7). The analysis revealed that the three stylistic attributes of color, photographs, and graphics are significantly related to overall satisfaction and account for 27% of the variation in satisfaction ratings. Four informational attributes were found to significantly affect satisfaction, accounting for 25% of the variation in ratings: a description of water treatment process, an FAQ section, the presence of informational maps, and a section on how to read the tables. All utilities, then, are urged to include these stylistic and informational features in their CCRs. See Chapter 7 for a full discussion of the survey and its implications for further interpretation of the attribute characterization.

One additional conclusion from the central site survey results is appropriate to discuss here—that the basic trust customers have in their own utility is significantly related to their overall satisfaction with a CCR. In other words, the most basic information contained in CCRs is likely to meet the needs of customers who already trust their utility. Those who are less trusting presumably have a greater need or desire for more information, and may be able to be influenced more strongly by the information contained in the CCR. Utilities should perhaps keep this group in mind when designing their CCRs. As presented in Chapter 3, almost 40% of the utilities stated that disseminating CCRs to their customers has improved customer relations, and 80% believe that CCRs are a good way to inform customers about drinking water quality. We recommend, then, that utilities take the time to produce high-quality CCRs that are attractive, easy to read, and contain more than the required information, such as descriptions of system improvements, conservation tips, explanations of source water issues, and sources for more information (e.g., Web site addresses). Also, utilities should carefully explain the origins and levels of various contaminants (and whether the levels meet standards) and the risks that these contaminants present. In terms of risk, utilities should be clear on exactly which populations are at risk, how they might be able to minimize their risk, and where they can find further information on the topic.

CHAPTER 6
NATIONAL SURVEY ON CCRs AND THEIR IMPACT

INTRODUCTION AND OVERVIEW

This chapter presents the results of a national survey conducted as part of this project to determine public awareness and perceptions of annual water quality reports (CCRs) and to help gauge their impact on consumer attitudes about drinking water and water utilities.

Survey Goals and Objectives

The primary objectives for the national phone survey were to:

- Use a statistically valid, random and representative sample of U.S. adults nationwide
- Determine how many people recalled getting a CCR
- Measure their reactions to and perceptions of that report
- Assess the impact of CCRs on customer perceptions of water quality, information, and local water utilities by:
  - Comparing these perceptions among customers who do and do not recall getting a CCR
  - Tracking changes in these perceptions since the CCR requirements were implemented in 1999

Other key goals for the project overall included identifying the characteristics, attributes, or conditions that make for more successful CCRs. Although this was best addressed by the central site survey in Chapter 7, we were also able to generate some relevant information and analysis as part of the national phone survey.

More specifically, the survey and analysis were designed to answer the following research questions (listed in order of their presentation in this chapter):

- How much information do people perceive they are receiving about drinking water safety? From their local water utility? From the media and other sources?
- How satisfied are people with the information they are getting? To what extent do they think they are receiving the information they need?
- To what extent are people aware of the CCRs? How many recall getting one from their local water utility? How many people have actually read them?
- What kind of impression have these reports made? How satisfied are people with the information and the way it is presented? How do they rate the reports in terms of being noticeable, attractive, informative, credible, and effective in making them feel more confident?
- Does receiving a CCR make a significant difference in their confidence and satisfaction? Are they more likely than other customers to be satisfied with the information they receive?

- What, if any, impact have the CCRs had on the attitudes of people who recall seeing them? Are they any more confident or satisfied than other customers? Do they have more positive perceptions of their tap water and local water utility?
- What, if any, impact have the CCRs had on long-term trends in consumer attitudes? To what extent have attitudes and perceptions changed over time—before and after CCRs were required nationwide? What other factors are behind these trends?
- What makes a CCR more *usable* for consumers and more *effective* for water utilities? How do customer perceptions of CCR characteristics affect their interest, satisfaction, and impact?

Survey Design and Methods

The survey was directed by Robert Hurd, PhD (Hurd & Associates, Inc.) and designed to ensure that the survey methods and results were comparable to other national phone surveys he conducted for AwwaRF in 1993 and 1997-1998. Many questions from previous surveys were repeated for tracking purposes. The survey was developed in consultation with other members of the research team, and then reviewed and approved by the AwwaRF project manager and Project Advisory Committee (PAC).

Survey topics included customer satisfaction, perceptions of water quality and safety, and awareness and perceptions of the information people received from their local water utilities, including the annual water quality reports called "Consumer Confidence Reports" (CCRs) required by the SDWA Amendments of 1998. The survey was also expanded to include additional questions to assess customer reactions to the wording of health effects language for arsenic, and explore in more detail issues related to point-of-use (POU) and filter devices.

After a successful "pre-test," the survey was administered by telephone to a random sample of 1,414 adult U.S. residents nationwide during December 2002. The interviews averaged between 8 and 15 minutes in length, depending on the type of respondent and the number of survey questions that applied to each group. The total weighted sample included 1,146 "customers" who get their tap water from a local water utility, plus 268 individuals who rely on their own private well.

Some efforts were made to identify and interview a somewhat larger number of actual customers (and fewer well users) than would otherwise be obtained in a random sample nationwide. This oversampling of customers had some impact on the demographics of the sample, but this was largely corrected through statistical weighting of the final sample (see discussion that follows).

The demographics of the survey sample compared favorably to known population parameters based on U.S. Census projections and previous AwwaRF surveys. However, the final sample was statistically weighted using standard industry techniques to correct for the oversampling of water utility customers; to make minor adjustments in some demographics (gender, age, and geographic region); and ensure comparability with previous surveys for tracking purposes.

Results from the total survey sample are accurate within ±2.6 percentage points, the margin of error allowed for samples of this size (n = 1,414) at the 95% confidence level. Samples based on subgroups have a somewhat larger margin of error, depending on their sample size. More detailed information on the survey sample and methods can be found in Appendix D.

INFORMATION ABOUT DRINKING WATER SAFETY

What People Have Seen, Read, or Heard in the Past Year

About three in ten water utility customers (28%) said they had seen, read, or heard something in the past year about their "local water utility or the quality and safety of the tap water it provides." The other seven in ten customers (70%) said they had *not* seen or heard anything, and 2% said they were "not sure."

Among those who said they had seen, read, or heard something (n = 163), about four in ten said it was "something positive" (40%). About another four in ten said it was "something negative" (41%). About two in ten said either "both" (9%) or it was "neither/neutral" (7%), and 2% were "not sure." Taken as a percentage of *all* water utility customers (not just those who had seen or heard something), this means that 11% of all customers have heard something positive and 12% have heard something negative about their local water utility or drinking water in the past year.

Local Concerns and Awareness of Threats

About two in ten customers (19%) said that they were aware of problems or incidents that had "hurt or threatened the quality and safety of the drinking water in their community" at one time or another. More than two in ten customers (25%) said that at one time or another they had seen, read, or heard something in the national or local media that "made them think the water in their own community might be unsafe or unhealthy to drink."

On the other hand, an eight in ten majority reported that they knew of no such problems, and had never heard anything in the media that made them worry about the quality and safety of their tap water.[18]

Amount of Information From Local Media and Water Utilities

Consumers are divided about the amount of information they are getting about drinking water safety, and the extent to which that information is meeting their needs.

About five in ten customers said that they get either "very little information" (35%) or "no information at all" (18%) from *the local news or media* about the quality and safety of their drinking water. On the other hand, more than four in ten reported that they get either "a lot of information" (11%) or "some information" (34%) on that topic from local media. One percent were not sure.

Table 6.1 shows that local water utilities did not fare any better than the media. More than four in ten customers said that they get "a lot" (13%) or "some" (31%) information about drinking water safety from *their local water utility*. But more than five in ten reported that they get either "very little" (29%) or "no information" (25%) from their water utility on that topic. Two percent were not sure. Further analysis (see Table 6.1) showed no significant difference in the amount of

[18] The results were the same for the total sample of all adults (including both water utility customers and private well users)—19% were aware of local water quality threats, and 25% said they had seen something in the media at some time that had made them worry about the safety of their own drinking water.

Table 6.1
Perceived amount of information obtained from local water utility versus local media

	Customers only		All adults	
Perceived amount of information	News/media (n = 1,146) (%)	Water utility (n = 1,146) (%)	News/media (n = 1,414) (%)	Water utility (n = 1,414) (%)
A lot of information (4)	11	13	10	11
Some information (3)	34	31	33	28
Very little information (2)	35	29	34	26
No information at all (1)	18	25	21	27
Not sure/no answer	1	2	2	8
Average rating (1-4)	2.4	2.3 ns	2.4	2.3*

*Statistically significant difference at the 0.01 probability level ($p < 0.01$) between sources (media versus utility).
"ns" denotes no statistically significant difference between sources (media versus utility).

information water utility customers perceived they were getting from the two sources—local news or media versus their local water utility.[19]

Meeting the Public's Needs for Information

Some customers are clearly getting the information they need. Others appear to be satisfied with less information, and some want more or better information on the quality and safety of their drinking water. Respondents were asked to indicate the extent to which they agreed or disagreed with the statement "I am getting the information I need…to know whether or not my tap water is safe to drink."

About three in ten customers said that they "strongly agreed" (31%) they were getting the information they need. More than 3 in 10 reported that they only "somewhat agreed" (37%). And three in ten answered that they either "somewhat disagreed" (17%) or "strongly disagreed" (13%) with that statement—indicating that they were not getting the information they need. Three percent were not sure.[20]

---

[19] Table 6.1 shows a significant difference, however, in the total sample of all adults (i.e., when private well users are also included in the sample). Adults in general perceived they were getting slightly more information from the local media (2.35 on the 1-4 scale) than directly from their local water utility (2.25), and this difference was statistically significant ($t = 3.69$, $df = 1286$, $p < 0.01$). This could be expected because private well users are not likely to get information from the local water utility because they are not customers.
[20] Similar results were obtained in the total sample of all adults (including both water utility customers and private well users). About three in ten adults "strongly agreed" (33%); more than three in ten only "somewhat agreed" (36%); and nearly three in ten either "somewhat disagreed" (15%) or "strongly disagreed" (12%)—indicating that they were not getting the information they need. Four percent were not sure.

Table 6.2
Satisfaction with information—amount versus type or quality

Level of satisfaction	Amount of information (n = 579) %	Type or quality of information (n = 579) %
Totally satisfied (5)	19	16
Mostly satisfied (4)	24	24
Somewhat satisfied (3)	24	24
Not very satisfied (2)	18	16
Not at all satisfied (1)	13	13
Not sure/no answer	2	8
Average rating (1-5)	3.2	3.2 ns

"ns" denotes no statistically significant difference between items.

Satisfaction With Utility Communications

Many customers are not fully satisfied with the information they are getting from their local water utilities, as shown in Table 6.2. Only four in ten customers were either "totally satisfied" (19%) or "mostly satisfied" (24%) with the amount of information they are getting about the quality and safety of their drinking water. And roughly the same number were either "totally satisfied" (16%) or "mostly satisfied" (24%) with the type or quality of information they are receiving.

Instead, more than half said they were only "somewhat" or less satisfied with either the amount or the type and quality of information they are getting from their local water utility about drinking water safety:

- About two in ten were only "somewhat satisfied" with the amount of information or with the type or quality of information they received
- About three in ten were either "not very" or "not at all" satisfied with the amount of information or with the type or quality of information they received

The extent to which local water utilities are meeting their customers' needs for information became even more clear when we cross tabulated two different survey questions—customer perceptions of the amount of information they are receiving information by their level of satisfaction with that amount. As Table 6.3 shows, this created the following subgroups and proportions:

- About 30% said they get at least "some" information (or "a lot"), and are either "totally" or "mostly" satisfied with that amount
- Another 11% are either "totally" or "mostly" satisfied, despite getting "little or no" information
- About 8% get only "some" information, and are only "somewhat" satisfied with that

Table 6.3
Subgroups based on amount of information and satisfaction

Level of satisfaction	Get a lot of information n %	Get some information n %	Little or no information n %	Not sure n %	Total n %
Totally or mostly satisfied	73 13%	99 17%	66 11%	10 2%	248 43%
Somewhat satisfied	1 < 1%	44 8%	93 16%	1 < 1%	139 24%
Not very or not at all satisfied	0 0%	10 2%	168 29%	0 0%	178 31%
Not sure	0 0%	1 < 1%	10 2%	3 1%	14 2%
Total	74 13%	154 27%	337 58%	14 2%	579 100%

- About 16% reported getting "little or no" information and are only "somewhat" satisfied
- About 29% said that they receive "little or no" information and are either "not very" or "not at all" satisfied" with that

Again, more than half of all customers said they were "somewhat" or less satisfied with their local water utility's communications, and this could be traced to the perceived amount of information they were receiving.

Perceived Improvement

Despite these unmet needs overall, some customers still gave their local water utility credit for recent improvements in customer communications. Although most customers said that their utility today was doing "about the same" (69%) compared to 5 years ago, about two in ten customers (19%) felt that it was doing a "better job" today of keeping them informed about drinking water quality and safety. Only 4% said their utility was doing "worse" and 8% were "not sure."

CCR AWARENESS, REACH, AND READERSHIP

Awareness and Recall of CCRs

About four in ten customers remembered getting a water quality report from their local water utility sometime in the past (38%). And about three in ten customers recalled receiving one *in the past year* (31%). Another two in ten customers were "not sure" if they had ever received a water quality report (18%), and about four in ten said that they had never received one (45%).

This current level of awareness is most likely to be considerably lower than the number of households that are actually sent CCRs. However, the figures above are important because they indicate the number of CCRs that were effective enough to be noticed in the first place and remembered up to a year or more later. This alone suggests there is considerable room for improvement. Demographic differences in the awareness (recall), reach, and readership of CCRs are presented and discussed later in this chapter under "Demographic Differences and the CCR."

Readership of the CCR

The vast majority of customers who reported getting a water quality report also said that they had read or skimmed through it personally, and many reported that other members of their household had read it. Among customers who recalled getting a CCR (n = 431), more than four in ten "read it thoroughly" (45%). Another two in ten read through "some of it" (18%) and three in ten just "skimmed through it quickly" (31%). Fewer than one in ten were "not sure" (1%) or "did not look at the report" (6%).

Altogether, more than 9 in 10 customers who got a report said they read or looked at it themselves (403 of the 431 who received a report, or 94%). Taken as a percentage of *all* water utility customers (not only those who recall getting a CCR), that means that the readership of water quality reports extends to at least 35% of all customers (403 of the 1,146 customers)—17% of all customers read the water quality report thoroughly, 7% read some of it, and 12% just skimmed through the report.

However, more than one person in the household often read the water quality reports. Among customers who received a CCR (n = 431), nearly four in ten said that someone else in their household read or looked at the report (35%). About one in ten were "not sure" (8%) and the other 56% said that no one else read the report.

Taken together, this means that the CCRs are read by someone in 95% of the households who reported getting one (409 of 431)—or by 36% of all customer households (409 of 1,146) nationwide. The vast majority of water utilities are required to distribute annual water quality reports, and the level of compliance is almost certain to be quite high. It is clear, then, that many customers are not aware of the CCRs they are getting—despite expressing a fairly high level of interest in that type of information. This suggests that the CCRs sent to some homes, for whatever reason, are not making a strong enough impression either to be noticed in the first place or remembered later.

Method of Delivery and Readership

Among the customers who recalled ever getting a CCR (n = 431), about four in ten said that it came along with their water bill (40%) and over four in ten said they received it as a separate mailing (46%). About 1 in 10 were "not sure" (12%) and 2% reported getting the report some "other" way.

Table 6.4 shows that customers who received their CCR as a separate mailing were significantly more likely to read the report and to read it thoroughly, compared to those who received their CCR along with their water bill. Customers who recalled getting their CCR as a separate mailing were also significantly more likely to say that someone else in the household read it (40%), compared with customers who reported getting it along with their water bill (28%). This difference was statistically significant at the 0.05 level ($p < 0.05$).

Table 6.4
Level of CCR reading by method of delivery

Reported level of CCR reading	Along with water bill (n = 173) %	As a separate mailing (n = 200) %
Read report thoroughly (4)	38	49
Read some of it (3)	17	20
Just skimmed through it (2)	34	28
Did not look at it (1)	10	3
Not sure/no answer	1	1
Average rating (1-4)	2.8	3.2*

*Statistically significant difference at the 0.01 probability level ($p < 0.01$) between delivery methods (separate mailing versus along with the water bill).

Note that the method of delivery has an important impact in determining which demographic groups are more likely to receive (or at least recall) a CCR. Most important is the tendency for "bill stuffers" to miss many consumers who rent and do not get a water bill directly from the utility. Separate mailings are more effective in reaching these renters, as well as other consumers as a whole. Again, these differences are discussed later in this chapter under "Demographic Differences and the CCR."

SATISFACTION AND PERCEPTIONS OF THE CCR

Satisfaction With the CCR

Most customers were fairly satisfied with the water quality report they received. Among customers who reported getting a CCR (n = 431), about seven in ten said they were either "totally satisfied" (35%) or "mostly satisfied" (38%) with "that report and the way it presented the information" to them. About two in ten were only "somewhat satisfied" (19%), and about one in ten were either "not very satisfied" (5%) or "not at all satisfied" (3%). Two percent were "not sure."

Perceptions of the CCR

If the water quality reports are to have the desired impact, they must first reach the customers and catch their attention. The reports need to stand out and be noticeable, they should look professional and attractive, they should be easy to read, they should be clear and understandable to consumers, they should be perceived as credible and believable, and they should present information that is important and worthwhile.

The national phone survey tested each of these perceptions by asking customers about the CCR they received and to indicate the extent to which they agreed or disagreed with a series of statements describing these characteristics. The results are presented in Table 6.5 and summarized in the sections that follow. A more detailed discussion of these characteristics and their impact can be found later in this chapter in the section titled "Creating a Usable and Effective CCR." Among

Table 6.5
Agreement with statements about and perceptions of the annual water quality report (weighted base: 406 customers who said they received a report and read or looked at it)

	Had important and worthwhile information %	Easy to read, clear, and understandable %	Stood out and got your attention %	Looked professional and attractive %	Seemed credible and believable %	Made you feel confident about safety %
Strongly agree (4)	57	58	42	61	54	53
Somewhat agree (3)	33	28	40	31	35	33
Somewhat disagree (2)	4	8	10	4	6	6
Strongly disagree (1)	4	4	5	1	3	6
Not sure	2	2	3	3	2	2
Average rating (1-4)	3.5	3.4	3.2	3.6	3.4	3.3

customers who reported that they received and either read or looked through a CCR (n = 406), the vast majority gave the reports positive ratings for content, appearance, presentation, and credibility.

- About nine in ten either "strongly agreed" (61%) or "somewhat agreed" (31%) that the report "looked professional and attractive." Only 5% disagreed.
- About nine in ten either "strongly" (57%) or "somewhat" (33%) agreed that the information was "important and worthwhile." Only 8% disagreed.
- About nine in ten either "strongly" (54%) or "somewhat" (35%) agreed that the report "seemed credible and believable." Only 9% disagreed.
- About eight in ten either "strongly" (58%) or "somewhat" (28%) agreed that the CCR was "easy to read, clear, and understandable." About 12% disagreed.
- About eight in ten either "strongly" (42%) or "somewhat" (40%) agreed that the report "stood out and got their attention." About 15% disagreed.

Note that consumers gave the water quality reports the lowest marks for standing out and getting their attention. Nevertheless, a majority of readers said the report was effective or successful in this regard.

However, it is also important to note that these ratings do not reflect customer perceptions of the full range of CCRs because, by necessity, the survey subsample here includes only those customers who recall getting a CCR, and therefore, only the CCRs that were already effective enough to be noticed in the first place and remembered up to a year or more later. And many of the attributes being measured are precisely the things that make CCRs more noticeable and memorable. A section later in this chapter titled "Creating a Usable and Effective CCR" analyzes the interrelationships among these ratings and other characteristics or perceptions, including the impact of these attributes on satisfaction with the CCR and consumer trust.

Perceived Impact of the CCR

Among customers who received a CCR (n = 431), more than four in ten said that getting the report made them feel either "a lot more confident" (22%) or "a little more confident" (23%). Fewer than one in ten said that it made them feel either "a little less confident" (4%) or "a lot less confident" (2%). More than four in ten (46%) said that getting the report made them feel "about the same," and 3% were "not sure."

This means that the net positive impact of the CCRs is 39% (45% minus 6%) among those who recall getting one. Taken as a percentage of all customers, that translates into a positive impact on 17% of all customers (197 of 1,146); a negative impact on only 2% of all customers (24 of 1,146); and a net positive impact of 15% (173 of 1,146, or 17% minus 2%).

Reported Content of the CCR

One reason why customers report that the CCRs made them feel more confident may be the fact that many utilities have no water quality or compliance problems to report. Among customers who received a CCR (n = 431), only 1% said that the report indicated "some major problems" with the quality and safety of their tap water. About two in ten (21%) said that the CCR reported "a few minor problems." And six in ten customers said that it reported "no problems at all" (61%). About two in ten were "not sure" what the report said. Taken as a percentage of *all* customers (including those who recalled getting a CCR, as well as those who did not), this means that 8% of all customers (97 of 1,146) recalled getting a CCR that reported any kind of problem at all, and less than 1% recalled any "major" problems.

Demographic Differences and the CCR

Further analysis revealed some important demographic differences in the reach (or recall), readership, and perceived impact of CCRs, as shown in Table 6.6. First, the table shows the results for all customers (n = 1,146)—31% recall getting a CCR in the past year, 38% recall getting one sometime in the past, 35% said that they read it, 30% agreed that the report made them confident their water was safe to drink, and 17% said that it made them feel "more confident" than before. Table 6.6 then shows similar breakdowns for demographic subgroups based on gender, age, household income, race, and the presence of children in the household. The table continues by showing breakdowns for subgroups based on homeownership (versus renting), geographic region, and type (or size) of community. Among the key findings:

- Homeowners were significantly "more likely" than renters to recall getting a CCR (46% versus 18% ever; 38% versus 13% in the past year); to report reading it (43% versus 17%); and to say that it either (1) made them confident that their water was safe to drink (37% versus 14%), or (2) made them more confident than before (21% versus 8%). These differences were statistically significant at the 0.01 level ($p < 0.01$), and may be one of the most important findings from the survey (see the discussion that follows).
- Age made a significant difference, with younger customers (age 18-29) being the least likely to recall getting a CCR, to report that they read it, or to say that it made them feel confident or more confident. These differences were statistically significant at the 0.01 level ($p < 0.01$).
- Household income also made a significant difference, with lower income households (under $20,000 per year) being the least likely to recall getting a CCR, to say that they had read it, or to say that it made them feel confident or more confident. Most of these differences were significant at the 0.01 level ($p < 0.01$).
- Significant differences were also detected based on region and on the reported type (or size) of community. The highest levels of recall, reading, and impact were reported in the Central and West regions (both west of the Mississippi River). The lowest levels were reported in the "central city" of major metropolitan areas.
- Nonwhites were *less likely* to recall getting a CCR, to say that they read it, or to report that it made them feel confident or more confident. But reading was the only difference that was statistically significant at the 0.05 level ($p < 0.05$).
- People with children in the household were *more likely* to recall getting a CCR, to report that they read it, and to say that it made them confident or more confident. But recall was the only difference that was statistically significant at the 0.05 level ($p < 0.05$).
- There were no significant differences based on gender. Men and women were equally likely to recall getting a CCR, to say that they read it, and to report that it made them confident or more confident.

Table 6.6
Demographic differences in CCR recall, reading, and perceived impact

Demographic group (n)	Recall CCR Past year %	 Ever %	Report reading the CCR %	CCR made you feel confident %	Even *more* confident than before %
All customers (n = 1,146)	31	38	35	30	17
Gender					
Men (531)	30	36	34	30	17
Women (615)	31	39	37	31	17
Age*					
18-29 years (257)	20	23	21	18	10
30-44 years (380)	33	39	36	30	17
45-59 years (256)	34	45	44	36	23
60+ years (227)	37	43	41	39	20
Household income*					
Under $20K (128)	19	23	22	20	13
$20-$40K (265)	27	36	32	27	16
$40-$70K (335)	33	41	38	34	19
Over $70K (275)	37	44	42	34	20
Race†					
White (870)	33	40	38	32	19
African-American (120)	23	27	25	21	13
All others (156)	24	32	30	26	12
Children in household†					
Yes (421)	35	41	38	33	19
No (725)	28	36	34	29	16
Own versus rent home‡					
Homeowner (810)	38	46	43	37	21
Renter (317)	31	39	37	14	8
Type of community‡					
Central city (175)	23	28	23	22	12
Metro suburbs (358)	35	43	35	36	23
Small town/city (453)	30	37	30	29	16
Rural area (145)	33	40	33	32	13
Household income‡					
East (or northeast) (230)	27	32	27	21	13
South (or southeast) (285)	29	34	29	30	17
Midwest (219)	25	30	25	27	15
Central (211)	38	49	38	41	22
West (Pacific) (201)	36	46	36	35	20

*Statistically significant difference at the 0.01 probability level ($p < 0.01$) between demographic subgroups based on age and household income (excluding the differences based on income for "made you feel confident.").
†Statistically significant difference at the 0.05 level ($p < 0.05$) between demographic subgroups based on race (for reading only), and based on children in the household (for recall of the CCR in the past year only).
‡Statistically significant difference at the 0.01 probability level ($p < 0.01$) between demographic subgroups based on homeowners (versus renters), type of community, and region (excluding the differences based on region for "made you more confident than before").

The difference between homeowners and renters may be one of the most important findings from the national phone survey because it:

- Accounts for and explains many of the other demographic differences noted above
- Is related to the method of delivery used by many utilities, which reduces the potential impact and effectiveness of CCRs

The survey sample included 1,414 adults nationwide, including 1,146 water utility customers and 268 who relied on their own private wells. The sample of water utility customers included 810 homeowners (71%) and 317 renters (28%). Only four in ten of these renters (40%, or 11% of all customers) reported getting a water bill from their local utility. Nearly six in ten renters (57%, or 16% of all customers) said that they do not receive a bill because water is included in their rent.

This means that water utilities who send their CCRs along with the water bill are likely to miss the 16% of all customers who rent and pay through their landlord. Relying on the landlord to distribute CCRs to tenants may not be effective. As noted earlier, customers were significantly more likely to recall getting a CCR if they received it as a separate mailing (versus with their water bill), and significantly more likely to say they (or someone else) read it and read it "thoroughly." This has a differential impact on various demographic subgroups that are more likely to be renters rather than homeowners.

In fact, the difference between homeowners and renters explains (or accounts for) many of the other demographic differences cited above. After controlling for this factor, many of the other differences in CCR recall and impact disappear, including:

- Most of the differences by age (except making you feel "confident")
- All of the differences by household income
- The difference in reading by race
- Most of the differences by community type (except making people feel "more confident" than before)

The only other differences not affected or explained by homeownership (versus renting) are the differences in CCR recall, reading, and impact by geographic region. Those continued to be statistically significant.

ASSESSING THE IMPACT OF CCRs

One of the key goals for the survey was to assess and document the possible impact that CCRs may be having on consumer attitudes, perceptions, and satisfaction. This was done in several ways: first, by asking consumers directly how it affected their confidence—as shown in the preceding section; second, by comparing to see if people who recalled getting a CCR were any more confident or satisfied than those who did not recall getting a water quality report; and finally, by tracking changes in customer attitudes before and after CCRs were required nationwide.

These results are presented and discussed in the sections that follow, first to show the impact of the CCRs on perceptions and satisfaction with information specifically, and then to show their impact more generally on satisfaction and perceptions of drinking water and local water utilities.

Table 6.7
Perceived amount of information from utility by CCR recall

Perceived amount from water utility	All customers (n = 1,146) %	Recall household ever getting a CCR? Recall CCR (n = 431) %	No recall (n = 515) %	Not sure (n = 200) %
A lot of information (4)	13	27	4	7
Some information (3)	31	43	19	31
Very little information (2)	29	24	32	31
No information at all (1)	25	5	43	25
Not sure/no answer	2	1	2	7
Average rating (1-4)	2.3	2.9*	1.8	2.2

*Statistically significant difference at the 1% probability level ($p < 0.01$) between customers who recalled getting a CCR and those who did not.

Impact on Perceptions and Satisfaction With Information

Further analysis showed that getting a CCR (or at least remembering it) had a significant impact on customer perceptions and satisfaction with the information they received from their local water utility.

Table 6.7 shows that customers who recalled getting a CCR (n = 431) perceived that they were getting more information from their local water utility compared to other customers, and these differences were statistically significant. Customers who recalled receiving a CCR were significantly more likely than others to credit their local water utility with furnishing "a lot of information" (27% versus 4%), "some information" (43% versus 19%), or any information at all (94% versus 55% total).

In addition, customers who recalled getting a CCR were also significantly more satisfied than other customers with both the amount of information from their water utility, as well as the type or quality of information they received.

Table 6.8 shows that among customers who recalled getting a CCR, more than six in ten were either "totally" or "mostly" satisfied with the amount of information they were getting. Among those who said they did not get a CCR, fewer than three in ten were totally or mostly satisfied. Instead, nearly half these customers were "not very" or "not at all" satisfied with the amount of information. These differences were statistically significant, and similar results were obtained for their satisfaction with the type or quality of information.

Clearly, getting a CCR (or remembering it) makes a significant difference in customer satisfaction with information.

Customers who recalled getting a CCR were also significantly more likely than other customers to agree that they were "getting the information they need…to know whether or not their tap water is safe to drink," as shown in Table 6.9.

Among customers who recalled getting a CCR, about eight in ten reported getting the information they needed. Only two in ten disagreed. Among those who said that they did not

Table 6.8
Impact of CCR on satisfaction with amount of information

Satisfaction with the amount of information received	All customers (n = 579) %	Recall household ever getting a CCR? Recall CCR (n = 209) %	No recall (n = 256) %	Not sure (n = 114) %
Totally satisfied (5)	19	33	9	16
Mostly satisfied (4)	24	31	19	21
Somewhat satisfied (3)	24	18	23	36
Not very satisfied (2)	18	11	25	16
Not at all satisfied (1)	13	7	20	9
Not sure	2	1	4	4
Average eating (1-5)	3.2	3.7*	2.7	3.2

*Statistically significant difference at the 1% probability level ($p < 0.01$) between customers who recalled getting a CCR and those who did not.

Table 6.9
Meeting information needs by CCR recall

Getting the information you need	All customers (n = 1,146) %	Recall household ever getting a CCR? Recall CCR (n = 431) %	No recall (n = 515) %	Not sure (n = 200) %
Strongly agree (4)	31	43	21	28
Somewhat agree (3)	37	37	35	43
Strongly agree (2)	17	11	21	17
Strongly agree (1)	13	7	19	10
Not sure	3	2	4	4
Average rating (1-4)	2.9	3.2*	2.6	2.9

*Statistically significant difference at the 1% probability level ($p < 0.01$) between customers who recalled getting a CCR and those who did not.

receive a CCR, about four in ten said that they were not getting the information they needed. This difference was statistically significant.

Finally, customers who recalled getting a CCR were significantly more likely (30%) than other customers (13%-15%) to feel that their local water utility was doing a "better job" today of keeping them informed, compared to 5 years earlier (see Table 6.10).

Impact of CCR on Perceptions of Water Quality

Nearly seven in ten customers rated the quality of their local tap water as either "excellent" (24%) or "good" (42%). About two in ten rated the quality as only "fair" (20%), and 13% said it

Table 6.10
Perceived changes in utility information by CCR recall

		Recall household ever getting a CCR?		
Perceived quality of tap water as	All customers (n = 579) %	Recall CCR (n = 209) %	No recall (n = 256) %	Not sure (n = 114) %
Better than 5 years before (3)	19	30	13	15
The same as 5 years before (2)	69	65	70	75
Worse that 5 years before (1)	4	2	7	1
Not sure	8	3	11	10
Average rating (1-3)	2.2	2.3*	2.1	2.2

*Statistically significant difference at the 1% probability level ($p < 0.01$) between customers who recalled getting a CCR and those who did not.

Table 6.11
Impact of CCR recall on water quality perceptions

		Recall household ever getting a CCR?		
Rated their water quality as	All customers (n = 1,146) %	Recall CCR (n = 431) %	No recall (n = 515) %	Not sure (n = 200) %
Excellent (4)	24	30	18	27
Good (3)	42	42	41	45
Fair (2)	20	20	21	17
Poor (1)	13	8	19	8
Not sure	1	1	2	3
Average rating (1-4)	2.8	2.9*	2.6	2.9

*Statistically significant difference at the 1% probability level ($p < 0.01$) between customers who recalled getting a CCR and those who did not.

was "poor." One percent were "not sure." When asked how the water quality today compared with 5 years earlier, most customers said it was "about the same" (68%). However, 16% said the water quality was "better" today, and only 10% said it was worse. Five percent were "not sure."

Tables 6.11 and 6.12 show that customers who recalled getting a CCR were significantly more likely to give their local tap water quality a positive rating, and to say that their local water utility was doing a "better job" today compared to 5 years earlier.

Table 6.11 shows that among customers who recalled getting a CCR, about seven in ten rated their water quality as "excellent" or "good," compared to only six in ten among those who said that they did not get a CCR. This difference was fairly modest, but statistically significant at the .05 probability level ($p < 0.05$).

Table 6.12
Perceived changes in water quality by CCR recall

		Recall household ever getting a CCR?		
	All customers (n = 579)	Recall CCR (n = 209)	No recall (n = 256)	Not sure (n = 114)
Perceived quality of tap water as	%	%	%	%
Better than 5 years before (3)	16	23	14	12
The same as 5 years before (2)	68	66	67	74
Worse than 5 years before (1)	10	8	13	6
Not sure	5	4	6	8
Average rating (1-3)	2.1	2.1*	2.0	2.1

*Statistically significant difference at the 5% probability level ($p < 0.05$) between customers who recalled getting a CCR and those who did not.

Table 6.12 shows that customers who recalled getting a CCR were significantly more likely (23%) than others (12%-14%) to say that their local water quality was "better" today than it was 5 years earlier. This was yet another modest, but statistically significant difference between the groups of customers. Getting a CCR (or at least remembering it) clearly makes a difference in customer perceptions of water quality and safety.

Perceptions of Safety, Standards, and the CCR

More than seven in ten customers said they thought the drinking water in their local community was either "better than the federal standards" (21%) or "met the federal standards" (52%). About one in ten said the water was "worse than the federal standards (12%), and nearly two in ten said that it either "depends" (1%) or they were "not sure" (14%).[21]

However, more than three in ten customers thought the federal standards are "not strict enough" (36%). Another 13% were "not sure" and 1% said "it depends." Roughly half believed that the federal standards are "about right" (49%) and only 2% said they were probably "too strict."[22]

These customer perceptions became more clear when the two questions were cross tabulated. Table 6.13 shows how this created the following subgroups and proportions:

[21] The results were similar in the total sample of all adults (including both water utility customers and private well users). More than seven in ten adults thought that the drinking water in their local community was either "better than the federal standards" (22%) or "met the federal standards" (51%). About one in ten said they thought the water was "worse than the federal standards (11%);two in ten said it either "depends" (1%) or they were "not sure" (16%).

[22] In the sample of all adults (including both customers and private well users), more than three in ten believed that the federal standards are "probably not strict enough" (35%); another 14% said they were "not sure" and 1% said "it depends." Less than half said the federal standards are "about right" (48%) and only 2% said they were probably "too strict."

Table 6.13
Subgroups based on perceptions of federal standards and local compliance

Perceive local tap water is	Standards too strict n %	Standards about right n %	Not strict enough n %	Depends/ not sure n %	Total n %
Better than standards	8 1%	145 13%	75 7%	15 1%	243 21%
Meets the standards	7 1%	329 29%	188 16%	77 7%	601 52%
Worse than standards	2 < 1%	27 2%	97 9%	6 1%	132 12%
Depends/ not sure	5 < 1%	55 5%	51 5%	59 5%	170 15%
Total	22 2%	556 49%	411 36%	157 14%	1,146 100%

Table 6.14
Perceptions of compliance with safety standards by CCR recall

		Recall household ever getting a CCR?		
Perceived their tap water as	All customers (n = 1,146) %	Recall CCR (n = 431) %	No recall (n = 515) %	Not sure (n = 200) %
Better than standards (3)	21	29	16	18
Meets the standards (2)	52	54	50	57
Worse than standards (1)	12	7	18	6
Not sure/depends	15	10	17	20
Average rating (1-3)	2.1	2.2*	2.0	2.2

*Statistically significant difference at the 1% probability level ($p < 0.01$) between customers who recalled getting a CCR and those who did not.

- About four in ten customers (43%) said their water meets the standards (or better) and the standards are adequate (or better)
- Another one in ten customers (8%) reported that their water meets the standards (or better), but they are not sure if the standards are strict enough
- About two in ten customers (23%) said their water meets the standards (or better), but perceive those standards are not strict enough
- About one in ten customers (12%) said their water is worse than the standards
- Another 15% said they were not sure how their water compared

This suggests that simply informing customers about compliance may not be enough to inspire confidence, particularly when a sizable minority is unsure about the adequacy of those standards. These results show that only four in ten customers (43%) have confidence in *both* the federal standards and local compliance.

Table 6.14 shows that customers who recall getting a water quality report from their local water utility were significantly more likely to say that their local drinking water either met or was better than the federal standards. People who did not recall getting a report were significantly less confident about local utility compliance. No significant differences were detected, however, in the perceived adequacy of those standards.

Tap Water Aesthetics and the CCR

The vast majority of water utility customers nationwide reported rarely or never having any problems with the taste, odor, or appearance of their tap water.[23] However, nearly two in ten customers said they "always" (13%) or "very often" (5%) have problems with the *taste* of their tap water. Another 15% "sometimes" have problems. Still, more than six in ten said they "rarely" (21%) or "never" (43%) have problems with taste. Three percent were "not sure."

Fewer than one in ten customers said they "always" (4%) or "very often" (4%) have problems with *odor* from their tap water. Another 14% said "sometimes." On the other hand, an eight in ten majority said they either "rarely" (23%) or "never" (56%) have problems with odor. One percent were "not sure."

Fewer than one in ten customers said that they "always" (3%) or "very often" (4%) have problems with the *color, clarity,* or *appearance* of their tap water. Another 15% said "sometimes." But an eight in ten majority reported they either "rarely" (23%) or "never" (53%) have problems with appearance. One percent were "not sure."

Further analysis suggests that the CCRs may also have some influence over these customer perceptions of taste, odor, and appearance, as shown in Table 6.15.

Customers who recalled getting a CCR were somewhat less likely (18%) than other customers (22%) to say that they "always" or "very often" have problems with the taste of their tap water. This difference was small, but statistically significant. Smaller (and nonsignificant) differences were found for odor and for the appearance of their tap water.

---

[23] Survey results for the sample of all adults (including both customers and private well users) were similar to those in the sample of customers only. Nearly two in ten adults said they "always" (12%) or "very often" (5%) have problems with the *taste* of their tap water. Another 14% "sometimes" have problems. Still, a seven in ten majority reported "rarely" (20%) or "never" (48%) having problems with taste. Three percent were "not sure."

Fewer than one in ten adults said that they "always" (4%) or "very often" (3%) have problems with *odor* from their tap water. Another 13% answered "sometimes." On the other hand, about eight in ten said they either "rarely" (21%) or "never" (59%) have problems with odor. One percent were "not sure."

Fewer than one in ten adults said they "always" (4%) or "very often" (4%) have problems with the color, clarity, or appearance of their tap water. Another 14% said "sometimes." But an eight in ten majority said that they either "rarely" (22%) or "never" (56%) have problems with appearance. One percent was "not sure."

Table 6.15
Frequency of problems with taste by CCR recall

Says they have problems with taste	All customers (n = 1,146) %	Recall household ever getting a CCR? Recall CCR (n = 431) %	No recall (n = 515) %	Not sure (n = 200) %
Always (1)	13	11	17	8
Very often (2)	5	7	5	4
Sometimes (3)	15	13	18	13
Rarely (4)	21	24	18	22
Never (5)	43	44	39	51
Not sure	3	2	3	4
Average rating (1-5)	3.8	3.9*	3.6	4.1

*Statistically significant difference at the 1% probability level ($p < 0.01$) between customers who recalled getting a CCR and those who did not.

Trust in Local Water Utilities and the CCR

About seven in ten customers said they trusted their local water utility either "totally" (26%) or "mostly" (44%). Another two in ten said they trusted their utility only "somewhat" (22%). Only one in ten said they trusted their utility either "not very much" (5%) or "not at all" (2%). About 1% were "not sure."[24]

Table 6.16 shows that customers who recalled getting a CCR were significantly more trusting of their utility than customers who did not recall getting one.

Overall Satisfaction, Perceptions, and the CCR

In addition to improving consumer confidence and perceptions of their local drinking water quality and safety, the CCRs appear to have a positive impact on overall satisfaction and on perceptions of their local water utility.

About seven in ten customers reported being either "totally satisfied" (38%) or "mostly satisfied" (32%) with their local water utility. About two in ten said they were "somewhat" satisfied (19%). Only one in ten were either "not very" (5%) or "not at all" (6%) satisfied. One percent were "not sure."

More than seven in ten customers rated their local water utility as doing either an "excellent job" (35%) or a "good job" (39%) overall. About two in ten customers felt that their

---

[24] Similar results were obtained in the total sample of all adults (including both water utility customers and private well users). More than six in ten adults said that they trusted their local water utility either "totally" (25%) or "mostly" (39%). Another two in ten said that they trusted their utility only "somewhat" (20%). Only one in ten reported trusting their utility either "not very much" (5%) or "not at all" (3%). About 2% were "not sure" and 7% said that it did not apply.

Table 6.16
Impact of CCR recall on trust in water utility

Trust their local water utility	All customers (n = 1,146) %	Recall household ever getting a CCR? Recall CCR (n = 431) %	No recall (n = 515) %	Not sure (n = 200) %
Totally (5)	26	28	25	27
Mostly (4)	44	49	37	50
Somewhat (3)	22	18	25	21
Not very much (2)	5	2	9	2
Not at all (1)	2	2	4	0
Not sure/no answer	1	1	< 1	2
Average rating (1-5)	3.9	4.0*	3.7	4.0

*Statistically significant difference at the 1% probability level ($p < 0.01$) between customers who recalled getting a CCR and those who did not.

Table 6.17
Overall satisfaction as a customer by CCR recall

Overall satisfaction as a customer	All customers (n = 1,146) %	Recall household ever getting a CCR? Recall CCR (n = 431) %	No recall (n = 515) %	Not sure (n = 200) %
Totally satisfied (5)	38	39	34	42
Mostly satisfied (4)	32	34	27	38
Somewhat satisfied (3)	19	19	20	15
Not very satisfied (2)	5	3	8	2
Not at all satisfied (1)	6	4	9	3
Not sure	1	< 1	2	1
Average rating (1-5)	3.9	4.0*	3.7	4.1

*Statistically significant difference at the 1% probability level ($p < 0.01$) between customers who recalled getting a CCR and those who did not.

utility was doing only a "fair" job (20%). Four percent said it was doing a "poor" job, and 2% were "not sure."

Tables 6.17 and 6.18 show that customers who recalled getting a CCR were significantly more satisfied than other customers with their local water utility and were significantly more likely to give it positive ratings for overall performance.

When asked how their local water utility was doing today compared with 5 five years ago, most customers said it was doing "about the same" (70%). However, 16% said their utility was doing "a better job" today, and only 6% said it was "worse" than 5 years earlier. About 8% were "not sure."

Table 6.18
Ratings of overall utility performance by CCR recall

Rated local water utility as doing (a/an)	All customers (n = 1,146) %	Recall household ever getting a CCR? Recall CCR (n = 431) %	No recall (n = 515) %	Not sure (n = 200) %
Excellent job (4)	35	39	30	39
Good job (3)	39	40	37	43
Fair job (2)	20	19	23	14
Poor job (1)	4	2	8	1
Not sure	2	1	2	3
Average rating (1-4)	3.1	3.2*	2.9	3.2

*Statistically significant difference at the 1% probability level ($p < 0.01$) between customers who recalled getting a CCR and those who did not.

Perceived Value and the CCR

Customers were also asked to rate the *value* of their drinking water "compared to other utilities like cable TV, telephone, gas, and electricity." About half said their water was an "excellent value" (18%) or "good value" (34%) for the price they paid compared to other utilities. About four in ten said the value for the price was only "fair" (25%) or "poor" (14%). One in ten were either "not sure" (4%) or said it did not apply (5%).

As we might expect from the other results, further analysis suggested that the CCR may also have an impact on the perceived value that customers assign to the product and service they receive from their local water utility. Table 6.19 shows that customers who recalled getting a CCR were significantly more likely (60%) than other customers (43%) to say their water utility provided a "good" or "excellent" value for the price they paid, compared to other local utility services. This difference was statistically significant.

Summary of CCR Impacts

Getting a CCR (or at least recalling receiving one) made a statistically significant difference in customer satisfaction and perceptions—including most aspects of the information, water quality, and overall performance delivered by the local water utility. More specifically, some important "gaps" or differentials were detected between customers who recalled getting a CCR and those who did not, including:

- Large and statistically significant gaps of 20 or more percentage points in perceptions of the amount of information received (43 points), satisfaction with that information (33 points), and the extent to which customers perceived their information needs have been met (20 points)

Table 6.19
Perceived value for price by CCR recall

Rated the value of their tap water as	All customers (n = 579) %	Recall household ever getting a CCR? Recall CCR (n = 209) %	No recall (n = 256) %	Not sure (n = 114) %
Excellent (4)	18	27	11	17
Good (3)	34	33	32	40
Fair (2)	25	21	29	25
Poor (1)	14	14	14	15
Not sure/no answer	9	5	14	3
Average rating (1-4)	2.6	2.8*	2.5	2.6

*Statistically significant difference at the 1% probability level ($p < 0.01$) between customers who recalled getting a CCR and those who did not.

- Moderate but significant gaps of 6 to 14 percentage points in perceptions of water quality (9 points), experiencing of actual problems with taste (6 points), and perceptions of local compliance with federal standards (14 points)
- Moderate but significant gaps of 7 to 13 percentage points in consumer trust (11 points), overall satisfaction as a customer (7 points), perceptions of overall utility performance (8 points), and perceptions of value for price (13 points)

In each case, customers who recalled getting a CCR gave their local water utility significantly higher ratings in these areas.

However, we should note that this positive impact was necessarily restricted only to those customers who recalled getting a CCR—38% of all water utility customers. Limitations in the current reach and awareness of CCRs may prevent them from having as much impact on customer satisfaction and perceptions as they otherwise could, as discussed in the section that follows.

TRENDS OVER TIME AND CCR IMPACT

Another way to assess the impact of the CCR is to track trends in consumer attitudes and perceptions over time and see how they have changed. To do this, we compared results from before and after CCRs were first required nationwide in 1999. Previous surveys conducted for AwwaRF by Dr. Hurd in 1993 and 1997-1998 provided an excellent baseline for making these comparisons (Hurd 1993, 1997, 1999, 2003). In designing the latest survey in 2002, he repeated many of the previous questions for tracking purposes.

These survey results are presented and compared here to show the extent of any changes from 1997-1998 to 2002 in (1) water utility communications; (2) customer awareness and perceptions of drinking water information; (3) perceptions of water quality, safety standards, compliance, and trust; and (4) overall satisfaction with and perceptions of water utility performance. This is followed by an overall assessment of CCR impacts on trends over time, and a discussion of economic conditions as a possible mitigating factor.

Changes in Water Utility Communications

In the 2 years before CCRs were required nationwide (1997 and 1998), about one in ten adults said they had received an annual water quality report (12% in 1997, and 13% in 1998). When noncustomers were excluded from the sample base, these numbers translate to about 15% of all water utility customers each year. These figures included consumers living in California (where similar reports had already been required for several years) and those served by water utilities that were already sending reports to customers voluntarily—in preparation for the new requirements or for other reasons.

Four years later in 2002, the number of customers who recalled ever getting a CCR has grown to nearly four in ten (38% of all customers; 31% in the past year). Although we can safely assume that far more households were sent CCRs, this figure shows the number of consumers who received a report that was usable or effective enough to be noticed in the first place and remembered up to a year or more later.

As noted earlier, many water utilities can probably meet the minimum regulatory requirements by simply sending information to their customers. But if everyone (both consumers and water utilities alike) is to realize the full benefit from these communications, it is important that the information be noticed, understood, and remembered by a much broader range of consumers.

Perceptions of Information Before and After CCRs

With the exception of the new CCRs, there appears to have been little change from 1997 and 1998 to 2002 in either (1) the amount of information that consumers say they are getting from local water utilities, the media, or other sources; or (2) in their perceptions of that information about drinking water quality and safety.

Customers perceive they are getting roughly the same amount of information (see Table 6.20) on drinking water quality *from their local water utilities* today—about four in ten reported getting "some" or "a lot" (42% in 1997-1998 and 44% in 2002).

However, they may also perceive that they are getting *a little less* information today from *the local media*—closing the gap that once existed between the two sources (see Hurd 2003). About five in ten customers (51%) reported getting "some" or "a lot" of information from the local media in 1997 and 1998, but that number dropped to 44% by 2002.

There has been little or no change in the number of customers who say that they have seen, read, or heard anything (positive or negative) in the past year about their local water utility or the quality of tap water it delivers (25% in 1997-1998 and 28% in 2002). There was also little or no change in customers' awareness of local incidents that threaten water quality (19% in 1997 and 19% in 2002).

However, a small increase was found in the number of customers who say that they have seen, read, or heard anything in the media that made them worry about the safety of their local drinking water—rising from 21% in 1997-1998 to 25% in 2002.

Perhaps most important, there has been no net change or improvement in the number of customers who feel that they are "getting the information they need to know...whether their tap

Table 6.20
Perceived amount of information from water utility—trends over time

Perceived amount of information	Customers only* 1997-1998 (n = 1,159) %	Customers only* 2002 (n = 1,146) %	All adults* 1997-1998 (n = 1,602) %	All adults* 2002 (n = 1,414) %
A lot of information (4)	11	13	10	11
Some information (3)	31	31	27	28
Very little information (2)	25	29	23	26
No information at all (1)	30	25	36	27
Not sure/no answer	3	2	4	8

*The total sample of all adults includes water utility customers plus households that rely on their own private wells for tap water. The "customer only" sample excludes private well users. tf

Table 6.21
Meeting information needs—trends over time

Getting the information you need	Customers only* 1997-1998 (n = 1,159) %	Customers only* 2002 (n = 1,146) %	All adults* 1997-1998 (n = 1,602) %	All adults* 2002 (n = 1,414) %
Strongly agree (4)	29	31	30	33
Somewhat agree (3)	40	37	38	36
Somewhat disagree (2)	16	17	16	15
Strongly disagree (1)	12	13	12	12
Not sure	3	3	4	4

*The total sample of all adults includes water utility customers plus households that rely on their own private wells for tap water. The "customer only" sample excludes private well users.

water was safe to drink." About seven in ten either "strongly agreed" (29% in 1997-1998 and 31% in 2002) or "somewhat agreed" (40% in 1997-1998 and 37% in 2002) with that statement, as shown in Table 6.21.

In addition, customers were no more likely in 2002 (19%) than in 1998 (18%) to say that their local water utility was doing a "better job" of keeping them informed than 5 years earlier. Most perceived that their local water utility was doing "about the same" (68% in 1998 and 69% in 2002). Fewer than one in ten said it was doing "worse" (4% in 1998 and 4% in 2002).

Taken as a whole, these results suggest that there has been little net change from 1997 and 1998 to 2002 in customer perceptions of the information they are receiving about drinking water safety—despite the growing awareness and recall of CCRs, and the impact they were shown to have earlier in this chapter.

Table 6.22
Perceptions of water quality—trends over time

	Customers only*		All adults*	
Rated the quality of local tap water as	1997-1998 (n = 1,159) %	2002 (n = 1,146) %	1997-1998 (n = 1,602) %	2002 (n = 1,414) %
Excellent (4)	24	24	25	25
Good (3)	43	42	42	42
Fair (2)	22	20	21	19
Poor (1)	10	13	10	12
Not sure	1	1	2	1

*The total sample of all adults includes water utility customers plus households that rely on their own private wells for tap water. The "customer only" sample excludes private well users.

However, it is also important to remember that the impact of CCRs applies only to the four in ten customers who can recall getting one. By not reaching more customers effectively, the potential impact of the CCRs is greatly reduced—this may help to explain why more changes in attitudes cannot be seen here.

Perceptions of Water Quality, Standards, Compliance, and Trust

A comparison of survey results from 1997-1998 and 2002 shows little or no change in customer perceptions of (1) water quality, (2) safety standards and local compliance, or (3) trust in local water utilities.

The preceding section of this chapter showed that customers who recalled getting a CCR gave their tap water significantly higher ratings for quality. But there was little change from 1997-1998 to 2002 in customer perceptions of their water quality, as shown in Table 6.22. Each year, nearly seven in ten rated the quality of their tap water as either "excellent" or "good" (67% in 1997-1998 and 66% in 2002). About two in ten rated the quality as only "fair," and one in ten thought that it was "poor."

Nor was there any change in the frequency with which customers said they experienced actual problems with the *taste*, odor, or appearance of their tap water. Each year, about six in ten said they either "rarely" or "never" have any problems with the taste of their tap water (63% in 1997-1998 and 64% in 2002). About 15% said that they had problems "sometimes." About two in ten each year reported "always" or "very often" having problems with taste (19% in 1997-1998 and 18% in 2002).[25]

[25] In addition, nearly eight in ten customers each year said that they either "rarely" or "never" have any problems with odor from their tap water (77% in 1997-1998 and 79% in 2002). About 14% said that they had problems "sometimes." About one in ten each year reported "always" or "very often" having problems with odor (9% in 1997-1998 and 8% in 2002).

And each year, more than seven in ten customers said that they either "rarely" or "never" have any problems with the color, clarity, or appearance of their tap water (74% in 1997-1998 and 76% in 2002). About 15% said that they had problems "sometimes." About one in ten each year reported "always" or "very often" having problems with appearance (10% in 1997-1998 and 7% in 2002).

Table 6.23
Trust in local water utility—trends over time

	Customers only*		All adults*	
Trust their local water utility	1997-1998 (n = 538) %	2002 (n = 1,146) %	1997-1998 (n = 800) %	2002 (n = 1,414) %
Totally (5)	27	26	26	25
Mostly (4)	43	44	37	39
Somewhat (3)	19	22	19	20
Not very much (2)	7	5	8	5
Not at all (1)	3	2	5	3
Not sure/no answer	1	1	5	9

*The total sample of all adults includes water utility customers plus households that rely on their own private wells for tap water. The "customer only" sample excludes private well users.

In 2002, customers were a little less likely than in 1997-1998 to say that their tap water was "better" than the federal standards (26% in 1997-1998 and 21% in 2002). But a seven in ten majority each year still agreed that their tap water either "meets the standards," (48% in 1997-1998 and 52% in 2002) or was better. And there was no change in the percentage of customers who said their water was "worse" than the federal standards (11% in 1997-1998 and 12% in 2002), or were "not sure" (16% in 1997-1998 and 15% in 2002).

There was no change in the *perceived adequacy* of the federal standards from 1998 to 2002. Nearly four in ten felt that the standards were "not strict enough" (37% in 1998 and 36% in 2002). About half believed that they were "about right" (47% in 1998 and 49% in 2002). Only 2%-3% said that the standards were "too strict," and 14% each year were "not sure."

Finally, Table 6.23 shows that there has been little change from 1998 to 2002 in consumer trust of local water utilities. Each year, about seven in ten customers said they trusted their local water utility either "totally" or "mostly" (70% in 1998 and 70% in 2002). About two in ten said that they trusted it "somewhat," and one in ten said they trusted it either "not very much" or "not at all" (10% in 1998 and 7% in 2002).

Overall Satisfaction and Utility Performance

There has been little, if any, change from 1997-1998 to 2002 in either overall "satisfaction as a customer" (see Table 6.24) or in perceptions of "overall utility performance" (see Table 6.25). This parallels the lack of change in customer perceptions of information, water quality, and trust—despite growing awareness and recall of CCRs.

About seven in ten water utility customers said they were either "totally" or "mostly" satisfied in 1998 (71%) and again in 2002 (70%), as shown in Table 6.24. About two in ten each year said they were only "somewhat" satisfied, and fewer than one in ten said they were "not very" or "not at all" satisfied (8% in 1998 and 11% in 2002).

Table 6.24
Overall satisfaction as a customer—trends over time

	Customers only*	
Overall satisfaction as a customer	1998 (n = 538) %	2002 (n = 1,146) %
Totally satisfied (5)	39	38
Mostly satisfied (4)	32	32
Somewhat satisfied (3)	19	19
Not very satisfied (2)	4	5
Not at all satisfied (1)	4	6
Not sure	1	1

*The total sample of all adults includes water utility customers plus households that rely on their own private wells. The "customer only" sample excludes private well users.

Table 6.25
Ratings of overall utility performance—trends over time

	Customers only*	
Rated their local water utility as doing (a/an)	1997-1998 (n = 1,159) %	2002 (n = 1,146) %
Excellent job (4)	34	35
Good job (3)	43	39
Fair job (2)	19	20
Poor job (1)	4	4
Not sure	1	2

*The total sample of all adults includes water utility customers plus households that rely on their own private wells. The "customer only" sample excludes private well users.

More than seven in ten customers said their local water utility was doing either an "excellent job" or a "good job" in each of the two surveys (77% in 1997-1998 and 74% in 2002). About two in ten rated their water utility's performance as only "fair" or "poor" each time (23% in 1997-1998 and 24% in 2002).

However, customers in 2002 were less likely than in 1998 to say their local water utility was doing a "better job" compared to 5 years earlier. That number dropped from 25% in 1998 to only 16% in 2002. Instead, customers were more likely to say that their utility was doing "about the same" (up from 58% in 1998 to 70% in 2002). But fewer than one in ten each year said their utility was doing a worse job (only 4% in 1998 and 6% in 2002).

Table 6.26
Perceived value for price—trends over time

	Customers only*	
Rated the value of their tap water as	1997-1998 (n = 1,159) %	2002 (n = 579) %
Excellent (4)	27	18
Good (3)	40	34
Fair (2)	23	25
Poor (1)	7	14
Not sure	4	9

*The total sample of all adults includes water utility customers plus households that rely on their own private wells. The "customer only" sample excludes private well users.

Changes in Perceived Value and Price and the Economy

There has been a significant drop in customer perceptions of the value of their tap water for the price they pay, as shown in Table 6.26. Nearly seven in ten customers (67%) in the 1997-1998 survey said their tap water was either an "excellent" or "good" value for the price compared to other local utility services. In the 2002 survey, only five in ten customers (52%) rated its value as either "excellent" or "good." Instead, about four in ten rated its value as only "fair" or "poor."

Given the lack of change in other customer perceptions, this drop in perceived value may be a reflection of some change in economic conditions from 1998 to 2002, as discussed in the sections that follow.

Customers in the 1998 and 2002 surveys were asked: "Compared to five years ago, would you say you and your family are better off today, worse off today, or about the same as five years ago?" About five in ten customers each year reported being "better off" (48% in 1998 and 51% in 2002), and nearly four in ten said that they were "about the same" (39% in 1998 and 35% in 2002). But there was a small increase in number who said they were "worse off"—rising from 8% in 1998 to 13% in 2002.

In either case, a drop in the perceived value of tap water would have a negative impact on overall customer satisfaction, and possibly on other perceptions. This may have offset some of the positive impact of the CCRs (discussed earlier in this chapter) and help to explain, at least in part, why we do not see more improvement in customer perceptions and satisfaction cannot be seen from 1997 and 1998 to 2002.

Overall Assessment of CCR Impacts

The earlier discussion was fairly conclusive in showing that customers who recalled getting a CCR were significantly more satisfied than other customers with the information, water quality, and overall performance they received from their local water utility. The gap or differential between those who did and did not recall getting a CCR ranged from 6 percentage points (for "taste") up to 14 percentage points (for local compliance)—making for a statistically significant difference in most cases. However, this impact or difference applied only to the 38% of all customers who could recall getting a CCR. The fact that more customers could not recall receiving a CCR limits the impact that the reports can have on the total population of all water utility customers nationwide. The 6 to 14 percentage point advantage is reduced to only 2 to 5 percentage points ($0.06 \times 0.38 = 2.2\%$ and $0.14 \times 0.38 = 5.3\%$) in the total sample of all customers.

Previous analyses of the 1997-1998 survey (Hurd 2003) traced an increase in overall customer satisfaction to a combination of small but statistically significant factors that included: (1) improved perceptions of water quality, safety standards, and trust; (2) better communications with customers of some water utilities; and (3) improved perceptions of price and value, along with economic conditions in general.

Our initial analysis here suggests that the same factors continue to drive overall customer satisfaction in 2002. But the results in this section indicate that some of these factors have changed (although others have not):

- Many more customers recall getting a CCR in 2002 than in 1997-1998 (38% versus 15%).
- There has been little or no change in the amount of information customers perceive they are getting, or in their perceptions of that information and how well it is meeting their needs.
- A small increase was found in the number of respondents who say they have seen, read, or heard something in the media that made them worry (21% versus 25%).
- Customer perceptions of their tap water quality, as well as the frequency with which they experience actual problems with taste, odor, or appearance, have changed little.
- Customers were a little less likely in 2002 than in 1997-1998 to say that their tap water is "better" than the federal standards, but the vast majority still agree that their water at least "meets" the standards. And, there was no change in the number who either said the water was "worse" or questioned whether the standards were adequate.
- Consumer trust of local water utilities has seen little or no change.
- In 2002, customers were a little more likely to say that their economic situation is worse than 5 years earlier (8% in 1997-1998 and 13% in 2002).
- They are much less likely to say their tap water is an "excellent" or "good" value for the price—a drop from 67% in 1997-1998 to only 52% in 2002.

Given these other trends, it may be that the small, but positive impact of the CCRs has been undercut or "masked" in 2002 by less positive perceptions of value for price and the economy in general. The net effect of this may be what we see here—little change in consumer attitudes, perceptions, and satisfaction from 1997-1998 to 2002.

Improvements in water utility communications, the economy, or both could then be expected to result in higher levels of consumer awareness, confidence, and satisfaction. The last section of this chapter focuses more specifically on what the survey indicates can be done to make CCRs and other communications more *usable* to the consumer and more *effective* for water utilities. This question also serves as the primary focus for the central site survey in Chapter 7.

CREATING A USABLE AND EFFECTIVE CCR

What Makes for a Usable and Effective CCR?

The primary goals of the national phone survey were to (1) determine how many people recalled getting a CCR; (2) measure their reactions to and perceptions of that report; and (3) assess the impact of CCRs on customer perceptions of water quality, information, and local water utilities. However, one of two key goals for the project overall was to also identify the characteristics, attributes, or conditions that make for more successful CCRs. Although this was best addressed by the central site survey in Chapter 7, we were also able to generate some relevant information and analysis as part of the national phone survey.

In this survey, we measured a number of different customer reactions to and perceptions of the CCRs that they received. Some of these perceptions corresponded with key concepts—*usability* and *effectiveness*—introduced in the literature review (see Chapter 2) and the analysis of CCR content and attributes (see Chapter 5).

For that reason, we were able to perform some additional analysis (described in the rest of this chapter) using the national survey data to help determine what makes a CCR more or less *usable* for consumers and *effective* for water utilities.

Indicators of Usability and Effectiveness

As noted earlier, customers were asked to indicate the extent to which they agreed or disagreed with a series of statements that described the CCRs and information they received as:

- Being important and worthwhile
- Easy to read and understand
- Standing out and getting their attention
- Looking professional and attractive
- Credible and believable
- Making them feel confident their tap water was safe to drink

Among those who recalled getting a CCR, at least eight in ten agreed with each of the statements—indicating that the vast majority had positive impressions.

Being "easy to read and understand" was most clearly related to the *usability* concept. Consumers want and need for their CCRs to be readable and understandable. This may also be a necessary condition for the CCR to be effective for water utilities.

Consumers also want and need for the information to be "credible and believable." Otherwise, the CCR is of limited use or benefit to either the consumer (usability) or the water utility (effectiveness).

Our literature review also indicated that consumers recognized the importance and value of having the CCR "stand out and get their attention." That would enhance both the usability for the consumer and the effectiveness for the water utility.

"Looking professional and attractive" might help in getting their attention, and thereby making the CCR more usable and effective. The same could be said for providing information perceived as "important and worthwhile."

An effective CCR for the water utility would make customers feel "confident" (or even *more* confident) that their water is safe and healthy to drink. To some extent that reassurance also meets consumer needs and desires—although it falls considerably short of the usability concept.

In addition to these six ratings, a separate question measured the perceived impact of the CCR by asking the extent to which it made them feel more or less confident than before they received the CCR.

Survey respondents were also asked how thoroughly they read the report (if at all), and how satisfied they were with the information and how it was presented.

Possible measures or indicators of usability could include the extent to which consumers can and do read the reports. Reading is also a likely precondition or indicator of its effectiveness for water utilities. Finally, satisfaction with the CCR would indicate that the consumer found it to be usable. And to some extent, it could also indicate that the CCR was (or would be) effective for the water utility.

Relationships Among CCR Characteristics

Our analysis showed that all of the agree/disagree statements were at least moderately correlated, ranging from r = 0.28 to r = 0.60 (where a correlation can range from -1.00 to +1.00). This indicates that of these perceptions are somewhat interrelated in the minds of consumers—people who give the CCR a positive rating on one characteristic are fairly likely to give it a positive rating on other characteristics.[26]

However, some of these relationships were considerably stronger than others. For example, CCRs that were perceived as "credible and believable" also strongly tended to make people feel "confident" that their water was safe and healthy to drink (r = 0.60). The following results present a fuller picture of these interrelationships.

CCRs rated highly for "standing out and getting attention" were somewhat more likely to be perceived as "professional and attractive" (r = 0.37) and "easy to read and understand" (r = 0.34). They also tended to be rated as somewhat more "important and worthwhile" (r = 0.33), a little more "credible and believable" (r = 0.30), and a little more likely to make people "confident" (r = 0.28).

CCRs rated highly for "looking professional and attractive" also tended to be rated as more "important and worthwhile" (r = 0.47), "credible and believable" (r = 0.40), and somewhat more likely to make people "confident" (r = 0.35). They also tended to get somewhat higher ratings for "standing out and getting attention" (r = 0.37), and for being "easy to read and understand" (r = 0.30).

CCRs rated as more "easy to read and understand" strongly tended to be perceived as more "important and worthwhile" (r = 0.55), more "credible and believable" (r = 0.40), and more likely to make people feel "confident" (r = 0.48). The relationship was less strong between "easy

[26] Correlations are a measure of the strength of relationship between two variables (i.e., measures or questions). They can range from -1.00 to 0.00 and from 0.00 to +1.00. A correlation of 0.00 indicates that there is no relationship at all between the two items—that changes in one item have no impact on the other. A correlation of +1.00 indicates the strongest positive relationship possible—a perfect one-to-one correspondence between the two items, where a change in one item would always be matched by a corresponding change in the other. A correlation of -1.00 indicates the strongest negative relationship possible—where increases in one item are always matched with corresponding decreases in the other item. For more information, see Hurd (2003).

to read and understand" and "looking professional and attractive" (r = 0.30) or "standing out and getting attention" (r = 0.34).

CCRs rated highly for containing information that is "important and worthwhile" tend to be those that are also rated highly for being "easy to read and understand" (r = 0.55) and "credible and believable" (r = 0.59), as well as for making people feel "confident" in their drinking water (r = 0.59). They also tend to be rated higher for "looking professional and attractive" (r = 0.47) and somewhat higher for "standing out and getting attention" (r = 0.37).

CCRs rated as more "credible and believable" strongly tended to be perceived as more "important and worthwhile" (r = 0.59) and more likely to make people "confident" (r = 0.60). They were also tended to get higher ratings for being "easy to read and understand" (r = 0.40), "looking professional and attractive" (r = 0.40), and "standing out and getting attention" (r = 0.30).

CCRs rated highly for making people feel "confident" in drinking water safety also strongly tended to be perceived as more "credible and believable" (r = 0.60) and "important and worthwhile" (r = 0.56). They were also more likely to be rated as easier to "read and understand" (r = 0.48), somewhat more "professional and attractive" (r = 0.35), and a little more likely to "stand out and get attention" (r = 0.28). A similar, but less pronounced, pattern was found for the separate question measuring the extent to which the CCR made people feel even "more confident" than before.

None of these CCR characteristics, however, were strongly related with how thoroughly customers said they had read the reports. CCRs rated highly for "standing out and getting attention" (r = 0.31) were somewhat more likely to be read thoroughly, as were those rated highly for being "easy to read and understand" (r = 0.21) and "important and worthwhile" (r = 0.24).

More thorough reading, however, did not necessarily translate into higher "satisfaction" (r = 0.23) or greater perceived impact (r = 0.22). Satisfaction with the CCR information and how it was presented was strongly related to making people feel "confident" about drinking water safety (r = 0.66) and being perceived as "important and worthwhile" (r = 0.62), "credible and believable" (r = 0.59), and "easy to read and understand" (r = 0.55). These characteristics had the most impact on satisfaction. Satisfaction was moderately related to CCR ratings for "standing out and getting attention" (r = 0.37) and "looking professional and attractive" (r = 0.37).

Further analysis showed that consumer trust in their local water utility was strongly related with the extent to which consumers rated their CCRs highly for making them feel "confident" (r = 0.65), providing "credible" (r = 0.49) and "important" (r = 0.51) information on drinking water safety, and being "easy to read and understand" (r = 0.39).

Trust was also related to seeing or hearing something positive (r = 0.47) about their local water utility, rather than hearing something negative, being aware of local water quality threats (r = -0.23) or hearing something in the media that made them worry (r = 0.25). Finally, there was also a strong tendency for consumers who trust their water utility to also be more satisfied with their CCRs, and vice versa. Those who were less satisfied with their CCRs strongly tended to be less trustful of their water utility, and vice versa.

Implications for Key Research Questions

Taken as a whole, the correlation results described above may have some important implications for the key questions in this project, including:

- What does it take for a CCR to "stand out" and get noticed? This question is critical to creating more usable and effective CCRs, and is addressed more thoroughly by the central site survey in Chapter 7.[27] The survey results here simply suggest that it helps to look "professional and attractive," provide "important" information, and to be "easy to read and understand." Some of what it takes for a CCR to stand out and get noticed may also depend on individual reader interest in the topic. It should be noted that CCRs received the lowest relative ratings for "standing out and getting attention," even though the subsample included only those who could remember the CCR they received. Improving this characteristic may be the key to increasing the reach, usability and effectiveness, and impact of CCRs.
- How important is it for the CCR to look "professional and attractive?" These results suggest that looking "professional and attractive" helps the material to stand out and demand a little more attention, and adds to the perceived credibility and importance of the information.

  This should not be overlooked because it tends to somewhat increase the likelihood that the report will be read and have the desired effect of making people feel "confident." Most consumers in our survey who remembered getting a CCR gave it high ratings for this characteristic.
- How important is it for the CCR to be "easy to read and understand?" This analysis shows that when reports are easy to read and understand, they are much more likely to be perceived as being "worthwhile" and to make people feel "confident" that their water is safe and healthy to drink. Most consumers in our survey who remembered getting a CCR gave it high ratings for this characteristic.
- What makes a CCR "credible and believable?" And how important is that? This analysis indicates that it helps to look "professional and attractive," to provide "important" information, and to be "easy to read and understand."

  The literature review suggests that this may also depend on other external factors like consumer trust, previous communications by the water utility, and its pre-existing relationship with consumers in that community.

  Credibility is critical, however, because it is one of the most important factors in determining whether the CCR will be effective in making customers feel "confident" about the safety of their drinking water.

  Among consumers in our survey who could remember getting a CCR, most gave the report high ratings for "credibility."
- What makes a CCR "important and worthwhile" to consumers? What makes them "satisfied" with these reports? The analysis suggests that professional and attractive

[27] This question is more appropriate for the central site survey because that task featured an experimental design in which participants were given two actual CCRs and asked to read them before answering a survey to measure their reactions and perceptions of each. This made it possible to give different CCRs to different participants, and to make direct comparisons in their reactions to different types of information and attributes (including format, style, and presentation).

In contrast, the national phone survey could ask people only about the CCRs they remembered getting. By definition, this restricted the sample to include only those consumers who could remember getting a CCR, and only those CCRs that "stood out" enough to be noticed in the first place and that made a strong enough impression to be remembered up to a year or more later.

materials are more likely to be perceived as credible and important, and that information on drinking water safety that is credible and easy to read or understand is much more likely to be perceived as worthwhile.

These same factors determine consumer satisfaction with CCRs. The most important factor, however, is the extent to which the CCR makes people feel "confident" about drinking water safety.

- What can water utilities do to get people to read the CCRs more thoroughly? And how important is that? Our analysis indicates that it helps when CCRs are perceived as "standing out and getting attention," being "easy to read and understand," and containing information that is "important and worthwhile."

  However, it also suggests that people who read their CCRs more thoroughly are not necessarily much more satisfied than those who read more casually, and the perceived impact on confidence is not that much greater either.

  It is important to note that our subsample includes only those people who recall getting a CCR and only those CCRs that were noticed in the first place and remembered up to a year or more afterward.

  If CCR materials are not noticed, they are neither *usable* for consumers nor *effective* for water utilities. If they are not easy to read and understand, they are less likely to be read at all, and more thorough reading is not likely to generate significant benefits for either the consumer or the utility.

  Again, the central site survey in Chapter 7 focuses more effectively on these questions as to what specific characteristics are most likely to result in consumer attention, involvement, understanding, and satisfaction.

- What does it take for a CCR to make people feel "confident" in their drinking water? This analysis shows that inspiring "confidence" is strongly related to providing information that consumers perceive as "important," "credible," and "easy to read and understand."

  Again, other external factors are likely to play an equally important role. These include consumer trust and a history of communications, outreach, and efforts by the water utility to build positive relations with consumers in that community.

  Among consumers in our survey who could remember getting a CCR, most gave the report high ratings for making them feel confident.

  Increasing consumer confidence is likely to depend on the ability of water utilities to communicate more effectively with customers who are not yet aware of the CCRs or the information they contain.

- What is the role of consumer trust? To what extent does trust influence satisfaction with the CCR, and/or to what extent can the CCRs help build trust? Our analysis indicates that consumer trust is central to customer satisfaction and the effectiveness of CCRs. Once it has been established over time, trust benefits water utilities, and CCRs can probably play a key role in that process.

- What is the relationship between usability and effectiveness? The results of this analysis show that there may be considerable overlap between the characteristics that make CCRs *usable* for consumers and *effective* for water utilities.

  The best indicators of usability were "satisfaction" and the extent to which consumers perceived their CCRs to be "easy to read and understand" and "important and worthwhile." However, these were most highly correlated with making people feel "confident" in drinking water safety—a key measure of effectiveness for water utilities.

This suggests that furnishing usable information can also help water utilities to achieve their goals of improving consumer confidence, perceptions, and satisfaction. And there may be little difference between the needs, desires, and interests of water utilities and their customers.

CONCLUSIONS

The survey suggests that CCRs are making a significant difference, but there is considerable improvement in making the materials more noticeable and using effective methods of delivery. Both water utilities and their customers could benefit from efforts and improvements in these areas. Although local water utilities send CCRs to the vast majority of U.S. households, the survey suggests that many of these CCRs have not been effective enough to be noticed by consumers in the first place or remembered up to a year or more later. Only 38% of all water utility customers recall ever getting a CCR (31% in the past year).

The survey also indicates that most customers are interested in the information and actually read the CCRs (as long as they know they received one). Our analysis showed that CCRs have a significant and positive impact on consumer confidence, perceptions and satisfaction with their local water utility. Compared to other customers, those who recalled receiving a CCR gave their local water utility significantly higher ratings for information, water quality, overall performance and customer satisfaction. But this impact is restricted to those customers who actually receive a CCR (or a CCR that is effective in standing out and getting their attention).

The survey also showed that the method of delivery makes a significant difference in awareness, reading and impact of CCRs. When reports are sent as separate mailings rather than "bill stuffers," they are more likely to reach renters (many of whom do not get a water bill directly). Separate mailings were also significantly more likely than bill stuffers to be recalled by consumers. In turn, that explains why homeowners are significantly more likely than renters to recall getting a CCR, reading it and reading it thoroughly, and feeling more confident in the safety of their drinking water as a result.

Our analysis showed that many of the same characteristics that make a CCR usable for the consumer also make them potentially more effective for the water utility. Chapter 7 examines these characteristics more thoroughly to determine how CCRs can be made more usable for consumers and effective for water utilities.

CHAPTER 7
A CENTRAL SITE SURVEY FOR EVALUATING CCRs

To estimate the relative impact of CCR attributes on customer evaluation of the reports, the project team implemented a central site survey. A central site survey, as the name implies, is conducted by having respondents come to a single location to respond to a questionnaire. This approach is used, for example, when respondents must react to objects or samples that cannot be disbursed to them where they live or work.

For this research, 152 people were recruited to read CCRs with known attributes (described in Chapter 5) and rate the reports on a number of dimensions. The relationship between attributes (the independent variables), and ratings (the dependent variables), were modeled using regression analysis. The models reveal which attributes are important in shaping a positive rating of the reports. The research indicates that the presence of several key stylistic and informational elements is significantly related to the overall rating of the report. Reports are likely to be highly rated and attract attention when they are colorful, make use of photographs and maps, and feature a section with FAQs. Tables should be kept simple; if they require a "how to read tables" section, they are too difficult or complex.

OBJECTIVES

The objective of the central site survey was to help utilities design effective CCRs by giving them a better understanding of which report attributes have an impact on consumer ratings. As indicated in Chapter 6, there is a significant positive impact on customers when they read their CCR. The research presented in this chapter allows us to examine what specific attributes of the CCR customers respond to and what characteristics of a CCR may encourage more customers to read their CCR.

SURVEY DEVELOPMENT

The utility survey, which was developed to be compatible in a number of ways with the mail survey portion of this research, was produced through an iterative process of design and review. It drew heavily from findings of the focus group research and the attribute characterization work. The survey, included in Appendix E, was also based, in part, on surveys developed for prior AwwaRF research projects.

SURVEY IMPLEMENTATION

For the central site survey of CCR ratings, 152 subjects were recruited to travel to a nearby research facility, read reports, and answer questions about them. Two sites were chosen for the research, Denver, Colo., and Brookfield, Wis., a suburb of Milwaukee. These sites were chosen because of their proximity to members of the research team, so that travel and related costs would be kept to a minimum. Although we assumed that regional differences in ratings of reports would be minimal, recruiting in two different locations allows at least a partial test of that assumption.

Subjects for the research were recruited by a professional research facility located at each site, from lists of prospective research participants that the facilities maintain. Participants were selected to achieve a balance of demographic characteristics, including gender, age, and education. Participants were required to receive their water from a water utility, but were not required to own the property where they live and receive tap water. This last decision was made because in some locations renters receive reports, and because the research team believed that the motivations inherent in the rating of CCRs are not inextricably associated with directly paying a bill. In other words, even people who pay for water indirectly through their rent have a potential interest in the quality of water they consume and would be motivated and able to rate CCRs. A survey question about home ownership was included to allow comparison of owners and nonowners.

At each location, three survey sessions were scheduled, for a total of six sessions. Between 28 and 33 participants were recruited for each session, so that at least 25 would attend. In all, 148 attended the sessions. Four additional surveys were completed in a pretest of the survey instrument conducted at the offices of Stratus Consulting, in Boulder, Colo. These responses have been included in the data analysis, bringing the total of survey respondents to 152.

Table 7.1 breaks down the respondents by key demographic variables.

In each session, participants first answered a series of questions about their own experiences with their water utility and about the quality of tap water they receive. Each participant then had 15 minutes to read one CCR and respond to questions about it. Each then had another 15 minutes to read a second CCR and respond to the same set of questions about that one. The survey concluded with a final set of demographic questions. Each participant rated two reports, so we gathered a total of 304 ratings.

A set of 100 CCRs, a subset of the 127 CCRs that had been received from a solicitation of water utilities around the country (see Chapter 3), was used in the research. To cull reports that were most redundant with respect to their characteristics from the set, the project team conducted a preliminary attribute analysis and pulled 27 reports. In this way, an even 100 reports were obtained for review. Two randomly selected reports were inserted into each of 50 envelopes. Half the envelopes were marked "A" and the other half were marked "B." Three of the six sessions were administered the "A" set of envelopes (two in Denver, one in Brookfield) and three sessions were administered the "B" set of envelopes (one in Denver, two in Brookfield).

Participants were informed of the project's purpose and told that a large number of CCRs were under review. They were asked to read the reports and to respond as if the reports had been sent directly to them. In itself, this instruction offers no assurance that participants were able to react in an altogether natural way. One might speculate, however, that there is a tendency for any decrease in the personal commitment individuals may have had to the task to be offset by the apparent importance they felt for the task as it was undertaken in a formal setting for which the subject was paid.

The success of this approach depends on the set of CCRs containing a cross section of levels on all attributes. An average of only three people rate any given report; however, each level of a two-level attribute, if it is evenly distributed across CCRs, will be observed 150 times. To discriminate among the relative influence of different attributes, there must be CCRs that comprise every combination. The reality, of course, is that attributes are not perfectly distributed across CCRs and that the presence of some attributes is correlated with the presence of others.

Table 7.1
Demographic attributes of central site survey respondents

Attribute	Category	Number (percentage)
Sex	Male	74 (48)
	Female	78 (51)
	No response	2 (1)
Age	Under 25	23 (15)
	25-34	29 (19)
	35-44	36 (23)
	45-54	26 (17)
	55-64	32 (21)
	65 and over	8 (5)
Education	Did not complete high school	1 (1)
	High school degree (or GED)	24 (16)
	Some college or technical school (no degree)	46 (30)
	Two-year college or technical school degree	15 (10)
	Four-year college degree	42 (27)
	Graduate work or degree	24 (16)
	No response	2 (1)
Race	White	113 (73)
	African-American	23 (15)
	Hispanic	13 (8)
	Asian Native American	1 (1)
	Other	1 (1)
	No response	2 (1)
Income	Under $20,000	10 (6)
	$20,000 to $39,999	33 (21)
	$40,000 to $69,999	55 (36)
	$70,000 or more	53 (34)
	No response	3 (2)

Table 7.2
"Overall, how satisfied are you with the report and the way it presents information to you?"

CCR rating	n	Percentage of responses (out of 301 responses)*
0 ("not at all satisfied")	10	3
10	16	5
20	11	4
30	12	4
40	14	5
50	25	8
60	18	6
70	39	13
80	56	19
90	63	21
100 ("completely satisfied")	37	12
Total	301	100

*Of 304 CCR ratings, 301 had a valid response for overall satisfaction.

CENTRAL SITE SURVEY RESULTS

Overall Satisfaction With CCRs

Participants rating the surveys were generally satisfied with the CCRs they reviewed. The mean rating on a 101-point overall satisfaction scale (0% = "not at all satisfied;" 100% = "completely satisfied") was 67.19.[28] Twenty-nine percent of CCR ratings registered 50% satisfaction or lower, and 48% of ratings were 70% satisfaction or lower. In the commercial sector, a 70% satisfaction rating is often used as a boundary between adequate and inadequate performance. By this standard, then, almost half the CCRs were judged inadequate. Table 7.2 shows the distribution of responses.

The survey participants also rated the CCRs on four specific dimensions—how useful is the information, how well organized is the information, how easy to read and understand is the language, and how attractive is the appearance. Again, there was general approval of reports on all these dimensions, with being well organized receiving the highest rating (mean 72% well organized) and appearance receiving the lowest rating (mean 63% attractive). Table 7.3 presents the number and proportion of report ratings above 70.

Although these findings show general satisfaction with the reports, they suggest that good organization and readability do not always yield a report that is considered useful.

[28] Functionally, the question is an 11-point scale because it is represented in 10-percentage-point increments.

Table 7.3
Satisfaction with four dimensions of CCRs

Dimension of satisfaction*	Number of report ratings above 70 (100-point scale; out of 301 responses)	Percentage of report ratings above 70 (100-point scale)
"How useful is the information?" (USEFUL)	146	49
"How well organized is the information?" (ORGANIZED)	177	59
"How easy to read and understand is the language?" (EASY-TO-READ1)	168	56
How attractive is the appearance?" (ATTRACTIVE)	133	44

*The word in capital letters following each question will be used in subsequent references to the item.

These items and the overall CCR rating were highly correlated.[29] Ratings of the attractiveness and readability of reports have somewhat lower correlations with other aspects; ratings of the usefulness and organization have somewhat higher correlations with other aspects. Nevertheless, even the lowest correlation of 0.63—between usefulness and readability—represents a substantial relationship. From this information it is not possible to determine whether readers are evaluating CCRs on a single dimension or on multiple dimensions. Table 7.4 is a matrix of correlation coefficients among the five measures of satisfaction with the CCRs.[30]

Participant Agreement With Descriptive Statements About the CCRs

For each CCR that the survey participants rated, a set of 12 statements was presented. Participants were asked whether they agreed or disagreed that the statements describe the report.

[29] Correlation is a measure of association between two variables. For this analysis, we used the statistic "Pearson's r," the value of which can vary between –1 and +1. A correlation of +1 or –1 between two satisfaction items would mean that knowledge of a satisfaction rating on one item provides perfect knowledge of satisfaction on the other. A correlation of 0 would mean knowledge of satisfaction on one item would provide no information about satisfaction on the other. For correlations that are not near 0, we can say that knowledge about one item tends to tell us something of the other, with the strength of the relationship greater as the correlation approaches –1 or +1. A positive sign indicates that the items tend to have high values or low values together; a negative coefficient indicates that when one item has high values, the other tends to have low values.

[30] The values correlated in this analysis, and other analyses in this report unless otherwise noted, are the *differences* between the rating given by each participant to the first and second CCRs. In effect, this represents a within-subjects analysis of the contrast between the two CCRs. Such an approach more accurately reflects the dependence of the two ratings on a single evaluative process and set of standards than does treating them as independent events. Also, analyzing the contrasts among the CCRs reduces the total variance resulting from differences among participants relative to the variance resulting from differences in reports.

Table 7.4
Correlations among CCR satisfaction ratings*

	Overall	Useful	Organized	Readable	Attractive
Overall	1.00				
Useful	0.85	1.00			
Well-organized	0.86	0.80	1.00		
Easy-to-read	0.78	0.63	0.71	1.00	
Attractive	0.84	0.69	0.73	0.64	1.00

*All correlations are significant ($p < 0.0001$).

The statements, which were derived from comments made about CCRs during the focus group research, represent different impressions a report might make on readers. In including these items, the intent was to help diagnose reasons for the overall satisfaction rating. A correlation between agreement with each statement and overall satisfaction indicates which impressions may be most important in determining satisfaction. Table 7.5 outlines the 12 statements, the percentage of ratings in agreement with the statement, and the correlation coefficient between the statement and overall satisfaction with the CCR.

Again, it is evident that a majority of participants rate the CCRs favorably on each statement, on average agreeing with positive statements and disagreeing with negative statements. Table 7.5 is sorted in the order of decreasing correlation with overall satisfaction. The most important impressions associated with a satisfactory rating of the reports are that it provides enough information, is clear, and seems professional. One statement, that the report leaves questions unanswered, is only quite weakly correlated with overall satisfaction.

It is interesting to note the majority of respondents (70%) agreed that the utility produces the report only because it is a requirement. This has a moderately strong negative correlation with overall satisfaction, suggesting that an effort to counteract this impression is worth the effort. Likewise, 60% of ratings agreed that the report could be mistaken for junk mail. This has a weaker correlation with overall satisfaction, suggesting that perhaps the item registers crowding in the respondents' mailbox as much as it does the appearance of the report itself. Nevertheless, the relationship is strong enough that utilities should develop ways of making their CCR stand out from the clutter of messages. See the sections that follow for an analysis of CCR attributes that attract attention.

It is important to understand that we cannot assume that a strong correlation between the statements and overall satisfaction means that the impressions *cause* overall satisfaction with the report. It could be that other kinds of impressions determine overall satisfaction, and that people who are rating the CCRs only justify or make sense of their satisfaction by agreeing with the statements. This possibility will be tested in the subsequent sections.

Table 7.5
Mean agreement with descriptive statements about CCRs, and correlation with overall CCR satisfaction

Descriptive statement: "This water quality report . . .*	Percentage of CCR ratings in agreement with statement	Correlation with overall satisfaction†
...gives enough information to decide whether to not the water is safe to drink." (ENOUGH INFORMATION)	80	0.77
...is easy to read, clear, and understandable." (EASY-TO-READ)	72	0.74
...is professional." (PROFESSIONAL)	82	0.74
...is credible and believable." (CREDIBLE)	87	0.69
...is designed more for public relations than to provide customers with real information." (PUBLIC RELATIONS)	45	-0.68
...is a waste of money." (WASTE OF MONEY)	22	-0.66
...would stand out and get your attention in the mail." (STANDS OUT)	60	0.65
...is something you would want to keep for future reference." (KEEP FOR FUTURE)	48	0.63
...was produced only because the utility was required to produce it." (ONLY REQUIRED)	70	-0.60
...is too technical to understand." (TOO TECHNICAL)	31	-0.59
...could be mistaken for junk mail." (JUNK MAIL)	60	-0.50
...leaves questions unanswered or gives only part of the story." (ONLY PART)	45	-0.19

*The word in capital letters following each question will be used in subsequent references to the item.
†All correlations are significant ($p < 0.0001$). Correlations are calculated on the contrast within subjects between the first and second report.

CCR Impact on Confidence

To evaluate the effect of CCRs on reader confidence, the survey included a question asking how the respondent felt about water quality and safety after reading the report. The findings from this item are shown in Table 7.6.

Sixty-three percent of CCR ratings induced readers to report either "a little" or "a lot" more confidence in the quality and safety of the water discussed in the report. Only 17% of ratings induced "a little" or "a lot" *less* confidence. It is important to realize that, among the reports evaluated in the study, the water actually was quite safe. In only one report did arsenic levels exceed the federal standard. No other violations were reported. There were, however, numerous discussions of contaminants that were present in the water at levels below the federal standard. Nevertheless, the research does not result in an effective test of the impact of reported standard violations on satisfaction.

Table 7.6
"If this were your own city, how would you feel about the quality and safety of your water after reading this report?"*

CCR rating	n	Percentage of responses (out of 300 responses)†
A lot more confident	87	29
A little more confident	101	34
Neither more nor less confident	64	21
A little less confident	29	10
A lot less confident	19	6
Total	300	100

*Future references to this item will use the label CONFIDENCE.
†Of 304 CCR ratings, 300 had a valid response for the effect on confidence.

The findings on confidence reactions to the CCRs must be interpreted with considerable caution. Although participants were asked to respond *as if* the report were for their own tap water, it should not be assumed that their involvement with the information was of the same intensity as if this were actually the case. Also, the responses were framed as a change in confidence relative to the respondents' current expectations.

Current satisfaction with drinking water quality does appear to have an impact on concern stimulated by the rated CCR. Among participants who rated their own water supply as "fair" or "poor," 53% of CCR ratings made the reader "a lot" or "a little" more confident.

Among participants who rated their own water supply as "excellent" or "good," 78% of CCR ratings made the reader "a lot" or "a little" more confident.[31] It appears that either respondents' experience with their own water supply had an impact on the way they interpreted the information in the reports, or a third factor caused both low ratings of water quality and the stimulation of concerns in reading the CCRs.

Dimensions of CCR Evaluation

The attribute analysis discussed in Chapter 5 anticipated four basic dimensions of evaluation: style, informational content, concern stimulus, and readability. Each was assumed to have some impact on the overall rating of the reports by customers. The results of the central site survey can be used to evaluate whether this scheme has any reality in the thought processes of customers themselves.

Relationships among the CCR rating questions can be investigated using the statistical method of *common factor analysis*. This method makes it possible to reduce a large set of survey items into a smaller set of more fundamental dimensions. The procedure analyzes correlations among survey items, locating clusters of items that are closely related, which define the factors.

Factor analysis is especially useful in analyzing attitudes because the psychological perspective that is assumed to be the basis of survey answers cannot be directly observed. Where

[31] The difference is significant. Continuity-adjusted chi-square = 15.16, df = 1, p = 0.0002.

Table 7.7
Factor pattern of CCR evaluation items

	Factor 1* (INFORMATIONAL VALUE)	Factor 2* (READABILITY)	Factor 3* (APPEARANCE)
ENOUGH INFORMATION	**0.85**	-0.11	-0.08
PUBLIC RELATIONS	**-0.79**	0.02	-0.04
USEFUL	**0.75**	-0.10	-0.11
CREDIBLE	**0.73**	0.03	-0.12
CONFIDENCE	**0.71**	0.03	-0.18
KEEP FOR FUTURE	**0.62**	-0.06	-0.11
PROFESSIONAL	**0.57**	-0.15	-0.45
WASTE OF MONEY	**-0.56**	0.17	0.07
WELL ORGANIZED	**0.55**	0.25	-0.19
OVERALL	0.44	-0.32	-0.33
ONLY REQUIRED	-0.40	0.12	0.22
JUNK MAIL	0.37	0.02	0.25
EASY-TO-READ2	0.23	**-0.68**	-0.01
TOO TECHNICAL	0.08	**0.90**	-0.05
EASY-TO-READ1	0.04	**0.87**	-0.09
STANDS OUT	0.25	-0.04	**-0.51**
ATTRACTIVE	0.01	-0.06	**-0.96**

*The words in capital letters following each factor will be used in subsequent references to the item.

survey items are highly correlated, that set of items is interpreted to be a composite measure of a single underlying attitude or perspective. In effect, the factor (i.e., attitude) is thought to generate the specific answers. For this analysis, the factors represent the different dimensions used by survey participants to evaluate the reports.

Seventeen variables were included in the factor analysis, including the 5 rating items, 11 of the 12 CCR statements discussed above, and the item measuring impact of the report on confidence.[32] The analysis identified three factors, given in Table 7.7.[33] The matrix is interpreted by looking across the rows to determine which factor has the highest coefficient for each item.

---

[32] The statement that the report "tells only part of the story" was excluded from the analysis because it had low correlation coefficients with all other variables.

[33] Maximum likelihood estimation was used to extract three factors using a promax oblique rotation. Although a number of items do not load well on any factor, solutions with four and five factors did not change the factor patterns of the first three factors and produced weak additional factors containing singletons (factors defined primarily by a single item). The values in the table are standardized regression coefficients, which, like correlation coefficients, vary from -1 to +1. Values near 0 indicate that the item is not important in identifying the factor. Values near -1 or +1 indicate that the item is important. The sign of the coefficient reflects the direction of the relationship.

Table 7.8
Correlations among rating factors

	Factor 1	Factor 2	Factor 3
Factor 1 (INFORMATIONAL VALUE)	1.00		
Factor 2 (READABILITY)	0.65	1.00	
Factor 3 (ATTRACTIVENESS)	0.68	0.59	1.00

The item is said to "load" on that factor. In an ideal solution, one factor will have a coefficient above ±0.50 and the other factors will have loadings near 0. (To help identify the underlying factor structure, loadings above 0.50 are represented in bold type.)

Factor 1 appears to identify the perceived informational value of the report. The items include those asserting the report has enough information and is useful, credible, professional, well organized, and something to keep for the future; and rejecting statements that it is produced only for public relations and is a waste of money. Factor 2 is defined by three items and clearly identifies a dimension related to the ease of reading and interpreting the report. Factor 3 is defined by two items, and identifies a dimension related to the appearance of the report. A number of items are not strongly related to any of the factors. Two of these items in effect trivialize the reports, either as junk mail or as a mere obligation that the utility is meeting. More interestingly, overall satisfaction is moderately related to all three factors. This suggests that the three underlying dimensions of interpretation are all related to overall satisfaction, although none of these predominates.

Three of the four composite attributes developed in Chapter 5, then, are confirmed to exist in the perspective of CCR respondents. The item measuring impact of the report on confidence (CONFIDENCE), however, loads on Factor 1 rather than on a factor of its own. For the attribute analysis, the project team expected CONFIDENCE to be inversely related to the presence of contaminant information. Instead it appears that readers rated their own sense of confidence based on the total amount of useful information within the reports. Once again, however, it is important to recall that only one of the reports contained a warning about a violation. In performing the factor analysis, the factors were allowed to correlate with one another. In other words, we assumed that the CCRs rated as presenting valuable information would tend to be the same ones that were readable and attractive. Table 7.8 presents the correlations among the three factors.

The implication of these correlations among factors is that all the dimensions combine to create an overall impression of a given CCR. Well-designed reports and those that are easy to read present information in an attractive way that makes it more accessible and useful. Although readers can distinguish among the different aspects of a report, the tendency is for utilities that do well on one aspect to do well on all.

Regression Modeling of Overall Satisfaction

In the final stage of analysis of the central site survey data, the research team looked for relationships among the subjective ratings made by customers and the relatively objective attribute characterization of reports, reported in Chapter 5. When utility managers understand

Table 7.9
Overall satisfaction as a function of style attributes

Dependent variable	Overall satisfaction	Parameter estimate	t-value	p value
	Y intercept	8.54	2.97	< 0.01
Independent variables	Color	14.58	2.45	0.02
	Photographs	15.09	2.91	< 0.01
	Graphics	8.99	1.90	0.06
	F value	19.37	(p < 0.01)	
	Adjusted R^2	0.27		

which CCR attributes are most closely associated with report satisfaction, they can better design the reports to meet customer needs.

For the main set of regression models, the survey item related to overall satisfaction with the CCR was used as the dependent variable. This seems to capture the most important concern to utilities; that is, the total impression that the CCR leaves. Four separate regression models were estimated, each using a set of attributes related to style, informational content, concern stimulus, or readability as independent variables. A stepwise regression procedure was used, which seeks out the subset of independent variables that meets statistical significance criteria and tends to produce a good overall fit of the data.

Model 1: Overall Satisfaction as a Function of Style Attributes

The first model predicts the level of overall satisfaction with the CCR based on seven attributes of style, including the layout, the use of color, the presence of photographs, the inclusion of graphics, the use of artistic maps, the amount of white space, and the font size used. These are all dummy variables, scored as "1" or "0" reflecting either the presence or the absence of an attribute, in the case of color, photos, graphics, and maps; or dividing the attribute into "high-style" (brochure layout, above-median white space, large font size) and "low-style" (newsletter layout, below-median white space, small font size) categories. This list of independent variables is a subset of the larger list of style attributes characterized earlier. Many of the other attributes showed little or no variation across the set of CCRs collected for this research, and thus would not be useful in predicting overall satisfaction.

Table 7.9 presents the results of the model. Three attributes—color, photographs, and graphics—are significantly related to overall satisfaction, accounting for 27% of the variation in ratings (i.e., the "Adjusted R^2").

The parameter estimates in the table indicate the impact of the attributes on overall satisfaction. We can see that using color in the reports is associated, on average, with 14.58 additional satisfaction points on the 101-point scale. Likewise, including photographs in the report is associated with 15.09 additional points above and beyond points associated with using color. The use of artistic graphics is associated with almost nine additional rating points.

Table 7.10
Overall satisfaction as a function of informational attributes

Dependent variable	Overall satisfaction	Parameter estimates	t-value	p value
	Y intercept	7.31	2.48	0.01
Independent variables	Water treatment description	18.36	3.75	< 0.01
	FAQs	20.26	3.12	< 0.01
	Maps	19.28	2.77	< 0.01
	How to read tables	-20.74	-3.44	< 0.01
	F value	14.24	($p < 0.01$)	
	Adjusted R^2	0.25		

Model 2: Overall Satisfaction as a Function of Informational Attributes

The second model predicts the level of overall satisfaction with the CCR based on nine informational attributes, including water source, treatment process, treatment standards, an FAQ section, informational maps,[34] more than one map present, an explanation of how to read the tables, a description of detected contaminants, and other information not categorized elsewhere. Again, these were scored as dummy variables ("1" or "0") and reflect the subset of informational attributes that has a degree of variance across the set of reports. Table 7.10 presents the results of the model. Four attributes—a description of the water treatment process, an FAQ section, the presence of informational maps, and a section on how to read the tables—are significantly related to overall satisfaction, accounting for 25% of the variation in ratings.

The parameter estimates in the table indicate the importance of the independent variables in predicting overall satisfaction. All are of similar importance. A discussion of the treatment process is associated with an additional 18.36 rating points on a 101-point scale; the inclusion of an FAQ section is associated with an additional 20.26 points; the use of informational maps is associated with 19.28 additional rating points, and the presence of a section on how to read the tables is associated with a *decrease* of 20.74 points.

The negative coefficient for the section on reading the tables is perplexing. An explanation may exist in the fact that the presence of such a section is also negatively correlated with ease of reading the reports ($r = 0.22$, $p < 0.01$). This feature of CCRs, then, is probably most commonly used when the text and tables are rather complex and difficult to understand. It appears that such a section does not compensate for clarity in the tables themselves.

Model 3: Overall Satisfaction as a Function of Concern Stimulus Attributes

The third model predicts the level of overall satisfaction with the CCR based on nine attributes related to the stimulus of concerns in readers, including a prominent statement that the water quality meets all standards, along with discussions of lead, copper, radon, *Cryptosporidium*,

[34] A few maps were clearly more decorative than informative, usually because they had very little detail and were used as visual interest rather than to convey information.

Table 7.11
Overall satisfaction as a function of concern stimulus attributes

Dependent variable	Overall satisfaction	Parameter estimates	t-value	p value
	Y intercept	7.76	2.54	0.01
Independent variables	*Cryptosporidium*	11.95	2.49	0.01
	Radon	27.59	3.91	0.01
	Lead	20.08	4.01	< 0.01
	Arsenic	26.16	2.16	0.03
	Other	-16.14	-2.59	0.01
	F value	7.22	(p < 0.01)	
	Adjusted R^2	0.17		

arsenic, trihalomethane, and other contaminants. Again, these were scored as dummy variables ("1" or "0") and reflect the subset of informational attributes that has a degree of variance across the set of reports.

Table 7.11 presents the results of the model. Five attributes—discussions of *Cryptosporidium*, radon, lead, arsenic, and other contaminants—are significantly related to overall satisfaction, accounting for 17% of the variation in ratings.

The parameter estimates in the table indicate that radon and arsenic discussions have the greatest importance in predicting overall satisfaction. Again, however, the sign of the coefficients is opposite of what was expected, because the presence of these discussions, and all others except "other contaminants," is associated with higher levels of satisfaction. Looking at correlations among these attributes and ratings of the reports, it becomes evident that discussions of contaminants do not in general alarm readers but instead reassure them that the reports contain enough information, are useful, and are not undertaken solely for public relations purposes or simply to meet the requirement.

Based on the focus group findings, the project team expected that a prominent statement assuring the reader that a utility's drinking water meets federal standards would be correlated with overall satisfaction with the report. This was not the case. Nor was the presence of a prominent reassuring statement correlated with the level of concern respondents expressed after reading the reports. Based on the central site survey, it appears this rather commonly used device for reassuring customers has no effect on satisfaction or confidence. In addition, if it discourages customers from reading the entire report, it may prove to be counterproductive. This is somewhat at odds with the findings from the focus group sessions, which were discussed in Chapter 4.

Model 4: Overall Satisfaction as a Function of Readability

The fourth model predicts the level of overall satisfaction with the CCR based on a single readability attribute, the Gunning Fog index, which assigns a reading difficulty level to text based on the length of words and sentences. The Gunning Fox index is a continuous variable.

No statistically significant association exists between the Gunning Fog index and overall satisfaction.

Table 7.12
Overall satisfaction as a function of three composite attributes

Dependent variable	Overall satisfaction	Standardized coefficient	t-value	p value
	Y intercept	0.00	2.57	0.01
Independent variables	Style	0.28	3.26	< 0.01
	Information	0.35	4.41	< 0.01
	Concern	0.12	1.60	0.11
	F value	28.62	(p < 0.01)	
	Adjusted R^2	0.35		

Because this was a significant surprise, a separate model was run with the rating related to ease of reading as the dependent variable and the Gunning Fog index as the independent variable. Again, no significant relationship was found. This does not contradict the finding of the factor analysis, reported previously, that readers do use ease of reading as a dimension in evaluating CCRs. It suggests only that the Gunning Fog Index does not adequately capture readability in a document as complex as these reports, where organization and format play a significant role in comprehensibility.

Model 5: Overall Satisfaction as a Function of all Significant Attributes

A final model was estimated using overall satisfaction as the dependent variable and three independent variables representing the three categories of attributes. Each independent variable was constructed as the sum of the values of the significant variables in the three previous models. Hence, (1) a variable "style" was created as the sum of "color," "photos," and "graphics"; (2) a variable "information" was created as the sum of "water treatment description," "FAQs," "maps," and "how to read tables" (reverse scored); and (3) a variable "concern" was created as the sum of "*Cryptosporidium,*" "radon," "lead," "arsenic," and "other."

Table 7.12 presents the results of the model. Only two of the composite attributes, style and information, are significantly related to overall satisfaction, although concern is almost significant. The model accounts for 35% of the variation in overall satisfaction.

Because the independent variables in this analysis are not dummy variables ("0" or "1") representing the presence or absence of an attribute, but instead scales produced by summing the values of other variables, the standardized coefficient is more useful to report than the parameter estimate. Because the standardized coefficient controls for the different metrics, it is not affected by the fact that different numbers of attributes were summed to arrive at the three independent variables. It shows the *relative* importance of the independent variables in predicting overall satisfaction.

From this model, we can conclude that the quality of information—in particular the items discussed in model 2—is the most important factor in determining overall satisfaction. The style of the report is nearly as important. The presence of a discussion of various contaminants is least important.

Table 7.13
Agreement that CCR stands out as a function of all attributes

Dependent variable	CCR stands out	Parameter estimate	t-value	p value
	Y intercept	7.61	3.01	< 0.01
Independent variables	Color	20.36	3.43	< 0.01
	Photographs	11.00	2.30	0.02
	FAQs	16.42	2.81	< 0.01
	Maps	13.88	2.16	0.03
	How to read tables	-20.11	-3.74	< 0.01
	White space	-13.51	3.08	< 0.01
	F value	28.97	(p < 0.01)	
	Adjusted R^2	0.56		

Regression Modeling of Attraction

One problem uncovered in the focus group research is that CCRs are often not seen or remembered when they arrive in the mail. If the reports are to have an impact on customer perceptions, they must first be read. Consequently, they must stand out and attract attention to themselves among the many messages that customers receive.

A regression model was estimated using as the dependent variable agreement or disagreement responses to the statement, "This water quality report would stand out and get your attention in the mail." For independent variables, all attributes used in the five models described earlier were included in a stepwise regression procedure that selects variables meeting a minimum significance criterion. The procedure selected six variables that are significantly related to standing out and getting attention. The model is quite powerful, explaining 56% of the variance in the dependent variable (Table 7.13).

Parameter estimates indicate that the use of color and photographs are important stylistic elements in attracting attention. Maps, which have both graphical and informational value, are also important. And an FAQ section attracts attention, possibly because it provides a place for readers to gather information about the utility and indicates an intention to respond to concerns. This feature may help avert the impression that the report is produced only because it is required. Once again, the presence of a section explaining how to read the tables is associated with reports that are less likely to stand out and attract attention. It may be that the presence of such a section signals that the text will be turgid and dull. A significant surprise in this model was that the presence of ample white space on the page, which creates a less dense appearance for the report, has a negative impact on the judgment that the report is likely to stand out.

CENTRAL SITE SURVEY CONCLUSIONS

The findings of this research suggest that utilities have been doing a reasonably good job of developing CCRs. Nevertheless, a number of lessons can be learned about the qualities of reports that satisfy customers and make the reports likely to stand out and be noticed. A remarkably consistent picture emerges:

- Reports should make use of color to draw attention and guide the reader through them.
- Reports should use photographs to improve the visual impact of the reports and provide a nonverbal channel of information to customers.
- Reports should make use of maps, again because they have visual appeal and present information in a way that complements the text.
- An FAQ section, prominently featured within the report, conveys information and signals the utility's willingness to communicate about customer interests and concerns. It may serve to avert the impression that the report is produced only because it is a requirement.
- A description of the water treatment process is considered valuable information by customers, and should be included in reports.
- Information tables should be kept as simple as possible. They should be self explanatory and not require interpretation or reading instructions.
- Utilities should not avoid discussion of contaminants found at measurable levels in the water. These are seen as supplying valuable information and do not appear to stimulate concerns.

Customers evaluate reports using at least three different criteria: informational content, readability, and appearance. The evidence here suggests that informational content is the most important of these, but that appearance is almost equally important in establishing overall satisfaction with the reports.

The attributes of CCRs that satisfy customers tend to be correlated with one another, suggesting two conclusions. First, synergies probably exist among the attributes, with good performance on one reinforcing and increasing others. Second, utilities that put considerable effort into their reports tend pay more attention to all attributes, and consequently produce higher quality documents that communicate better with their customers.

CHAPTER 8
CONCLUSIONS AND RECOMMENDATIONS

OBJECTIVES AND METHODS

The reauthorized SDWA of 1996 added provisions requiring all CWS to deliver annual water quality reports called Consumer Confidence Reports to their customers. With an estimated 46,000 CWS required to develop and distribute CCRs to their customers each year, the CCR requirement poses an important challenge for water utilities to disseminate information to their customers. The research reported here was conducted to evaluate the effectiveness of CCRs and to better understand their impact on water utility customers.

Using a variety of research methods, the project team qualitatively and quantitatively evaluated (1) whether and how CCRs have influenced consumer perceptions, and (2) what attributes of a utility communication effort have the most influence on consumer perceptions.

Our findings from the research conducted for this project and from a review of the existing literature reveal two fundamental "take-home" messages:

- CCRs can and do have a significant and positive impact on customers' perceptions of, attitudes toward, satisfaction with, and trust in their water utility
- In general, there is room for improving CCRs to communicate more effectively with utility customers

Depending on the objectives of and the resources available to the water utility, decision makers should consider the methods and findings of this research to explore approaches for improving the use of their own CCRs. The CCR can be a key element in communicating with water utility customers and for improving customer relations.

Starting with a review of what is currently known about public perceptions of drinking water safety, the impact of water utility communications on those perceptions, and the research methods used to make those assessments, we used several different techniques to conduct primary research aimed at a better understanding of CCR effectiveness. Table 8.1 outlines the research tasks, the general research question addressed by each task, the methods we used, and the number of respondents for each task.

RESULTS

The utility survey indicated that water utilities have a very wide range of experience with the development and implementation of their CCRs—often dependent on the size of the utility and the number of customers—a result that is probably a function of the resources available to water utilities for the customer relations efforts. Although there is no overwhelming indication that utilities consider the effort either a waste of time or an extremely valuable exercise, it is apparent that the CCRs require a significant investment on the part of utilities of all sizes (in terms of effort and cost).

When asked about their perceptions of the effects of CCRs on various water quality and customer communication issues, utilities generally agreed that CCRs are a good way to inform customers about drinking water quality, although not the only method that utilities should use to

Table 8.1
Research tasks, questions, methods, and respondents

Task	Research question	Methods	Respondents
Utility survey (reported on in Chapter 3)	What are water utilities doing to meet the CCR requirements?	Mail survey of water utilities	118 water utility respondents
	How do customers react as they read their CCRs?	Directed focus groups with water utility customers	10 focus groups in 5 cities
CCR attribute characterization (reported on in Chapter 5)	What attributes are most important in determining the usability of CCRs?	Coding CCRs on 95 attributes describing usability and effectiveness	127 CCRs coded
National phone survey (reported on in Chapter 6)	What is the current reach and impact of CCRs?	National random sample of general population via telephone survey	1,414 (1,146 water utility customers, 268 well water users)
Central site survey (reported on in Chapter 7)	How do customers react to CCR attributes?	Experimental implementation of survey evaluating CCRs on multiple response scales with water utility customers	152 subjects in three cities (Denver and Boulder, Colo., and Brookfield, Wis.)

Table 8.2
Extrapolating to the U.S. population

<table>
<tr><td colspan="5">104,705,000 U.S. households*</td></tr>
<tr><td colspan="5">78.2% on CWS (the remainder on well water)†
81,879,000 households served by CWS</td></tr>
<tr><td colspan="4">38% remember getting a CCR</td><td>62% do not remember getting a CCR</td></tr>
<tr><td colspan="4">31,114,000 households remember getting a CCR</td><td rowspan="4">50,765,000 households do not remember getting a CCR</td></tr>
<tr><td>45% read it thoroughly</td><td>18% read some of it</td><td>31% skimmed it</td><td>6% didn't look at it</td></tr>
<tr><td>14,001,000 households</td><td>5,601,000 households</td><td>9,645,000 households</td><td>1,866,840 households</td></tr>
<tr><td colspan="4">35% said someone else also looked at the CCR</td></tr>
</table>

*104,705,000 households as of March 2000. *Source:* U.S. Census, 2000.
†21.8% of the national telephone respondents were on well water (78.2% are served by CWS).

keep their customers informed. Nevertheless, the utilities were evenly divided about whether the CCRs had improved their customers' perceptions of the utility and drinking water quality. Similarly, only 38% of the sample utilities agreed that CCRs help to improve relationships with customers. Contradicting this somewhat skeptical perception from water utilities, findings from the national telephone survey of customers indicate that CCRs do have a significant and positive impact on their perceptions of the utility and drinking water quality. The analysis reveals a net positive impact of 39% in improving confidence in the health and safety of customers' tap water.

The national telephone survey asked participants if they remembered getting a CCR, and, if they remembered receiving it, if and how closely they read it. Table 8.2 shows how the answers to these questions translate to the U.S. population.

With about 78% of the 105 million U.S. households receiving their water from CWS, more than 81 million U.S. households could or should receive a CCR. With 38% of these remembering receiving a CCR in the past (31% remembering this within the last year), there are roughly 31 million households that would remember getting a CCR. Of these, 14.0 million read it closely, 5.6 million read some of it, and almost 9.7 million skimmed it. Thus, a total of 29.2 million households would have been "reached" by CCRs. Of these, in 35% of households, more than one person read the CCR. Perhaps 39.4 million people in the United States, then, would have read a CCR from their water utility (roughly 14.0% of the U.S. population or 13.3% of those served by a CWS).

Of customers who remembered receiving a CCR, 45% reported that getting the report made them feel a lot or a little more confident about the health and safety of their tap water. About 6% said that it made them feel a little or a lot less confident. This net positive impact of about four in ten customers translates to CCRs having a net positive impact on about 12.1 million U.S. households—improving their confidence about the health and safety of their tap water.

The impact a CCR will have on its readers depends on (1) the quality of the water and (2) how the information is presented in the CCR. The CCR attribute characterization illustrated that, although all CCRs must contain common basic information about water quality, a wide array of attributes can constitute a CCR. These attributes, in turn, affect the usability and effectiveness of CCRs. Our analysis of 127 sample CCRs revealed room for improving the appearance of CCRs to make them more attractive to utility customers.

The attribute analysis revealed that the three stylistic attributes of color, photographs, and graphics are significantly related to overall satisfaction, accounting for a significant portion of the variation in satisfaction ratings. Four informational attributes were found to significantly affect satisfaction: a description of water treatment process, an FAQ section, the presence of informational maps, and a how-to-read-tables section. Form and content of CCRs have a significant impact on customers' satisfaction with their water utilities.

In one important conclusion from the central site survey results, the basic trust customers have in their own utility was found to be significantly related to their overall satisfaction with a CCR. In other words, the most basic information contained in CCRs is likely to meet the needs of customers who already trust their utility. Those who are less trusting may have a greater need or desire for more information, and may be able to be influenced more strongly by the information that the CCR contains.

RECOMMENDATIONS

The findings of this research suggest that utilities have been doing a reasonably good job of developing CCRs. Nevertheless, a number of lessons can be learned about the qualities of reports that satisfy customers and make the reports more likely to stand out and be noticed. Customers evaluate reports using criteria such as informational content, readability, and appearance. The research suggests that informational content is the most important of these, but that appearance is almost equally important in establishing overall satisfaction with the reports. Given that utilities are required to implement CCRs and that there is a wide range of ways of doing so, much of the research in the project focused on whether and how customers read the CCRs they received and on what aspects of a CCR will elicit a response in customers. Through the microfocus group work, the research team postulated a three-phase model of CCR reading:

1. Initial sort—customers conduct a preliminary sorting process in which they discard what they believe to be "junk" or otherwise irrelevant mail
2. Scanning—customers will give the CCR a cursory scan, looking for red flags or prominent reassurances that everything is basically okay
3. Reading—if a CCR catches a customer's interest, he or she may read it more thoroughly and attempt to understand the information that it contains

Using this conceptual model, utility personnel responsible for developing a CCR can consider how different efforts affect different aspects of customers looking at their CCRs. Some efforts can be focused on getting the CCR past the initial sort. Some efforts may be allocated to improving the efficiency of information that individuals would read if they only scan the document. And other efforts can focus on improving the presentation of information for those individuals who get as far as reading the CCR.

Based on our conceptual three-phase model of CCR reading, we make the following recommendations for how water utilities may consider improving their CCRs to make them more effective at each stage of this process.

Initial Sort

With only 38% of customers remembering receiving a CCR, increasing the number of individuals who read their CCR has the greatest potential for having a net positive impact on utility customers. If they are to pass the initial sort, our results suggest that CCRs should look "professional and attractive," contain "important" information, and be "easy to read and understand." Some of what it takes for a CCR to stand out and get noticed may also depend on individual reader interest in the topic. Improving the "standing out and getting attention" characteristics of CCRs may be the key to increasing the reach, usability and effectiveness, and impact of CCRs. Some of the approaches to increasing readership that we have discussed in the various parts of this research include:

- Make the CCR look professional and distinct so that it stands out from junk mail. Prominently displaying the utility's name may help in this regard.
- Use a separate mailing rather than a bill stuffer. Bill stuffers reach only those who pay the bills, which misses a large portion of the population. In addition, they tend to be less noticed than a separate mailing.
- Distribute the CCRs to renters and other groups such as schools, nursing homes, and businesses (sending the CCRs only to property owners misses a significant portion of the population).
- Expend more effort to reach specific populations such as those living in central cities.

Skimming

Of those who remembered getting a CCR, only 45% said they read it closely. This suggests that CCRs could be designed to (1) provide important information that is readily accessible for those who will just skim it and (2) include information that would encourage people to read it more carefully. Some approaches to supplying this information and increasing the visual appeal of CCRs that we discussed in this research include:

- Prominently place any "take away" messages or statements expressing overall quality and safety, such as "Your tap water is even safer than the state and federal standards for drinking water require."
- Highlight issues of concern to sensitive subpopulations with any such prominent take away messages.
- Use color to draw attention to and guide the reader through the CCR. Use photographs to improve the visual impact of the CCR and to open a channel of nonverbal information with customers.
- Include maps, again because they have visual appeal and present information in a way that complements the text.

Reading

Our analysis showed that when reports are easy to read and understand, they are much more likely to be perceived as being "worthwhile" and to make people feel "confident" that their water is safe and healthy to drink. Analysis suggested that professional and attractive materials are more likely to be perceived as credible and important. In addition information on drinking water safety that is credible and easy to read or understand is much more likely to be perceived as worthwhile. These same factors determine consumer satisfaction with CCRs. The most important factor, however, is the extent to which the CCR makes people feel "confident" about drinking water safety. Our analysis shows that inspiring confidence is strongly related to disseminating information that consumers perceive as "important," "credible," and "easy to read and understand."

For individuals willing to make the commitment to reading the CCR more thoroughly, the information should be presented in a credible, accessible, and understandable format. In general, the objective should not be to *persuade* individuals not to worry about their water. Instead, the objective should be to *inform* individuals to improve their understanding of the water quality issues in their community. Those interested in reading the CCR closely may be doing so because they have serious concerns, meaning that they need to be informed and not convinced. An unbiased and open communication with customers will engender trust. Some approaches to improving information communication that the project team discussed in this research include:

- Including a short but prominent listing of contents very near the beginning or initial point of focus, possibly even on the cover
- Offering an FAQ section or a similarly abbreviated section that allows the customer to quickly locate a particular topic of interest
- Supplying contact information for individuals who want more information (even for those who do not want more information, including the contact information indicates that the utility is open to communication, which engenders trust and satisfaction)
- Describing the water treatment process, which customers consider to be valuable information
- Keeping information tables as simple as possible; they should be self explanatory and not require interpretation or reading instructions
- Not avoiding a discussion of contaminants found at measurable levels in the water—such discussions are seen as supplying valuable information and do not appear to stimulate unwarranted concerns

Every water utility has somewhat different objectives with respect to their customer communication efforts and every water utility has their own experience and history with developing and implementing their CCRs. One assumption in undertaking this research has been that utilities are interested in improving customer relations and increasing customer satisfaction with the water utility. While not all of the recommendations outlined above will be applicable in all settings to achieve improved customer satisfaction, the fact that CCRs can and do play a significant role in influencing customer attitudes toward the utility, suggests that many utilities can still benefit from taking a more proactive role in designing, evaluating, and implementing their CCRs. It is hoped that the methods used in this research and the results of our analysis will offer guidance in the quest for more effective CCRs.

APPENDIX A
UTILITY SURVEY

YOUR WATER UTILITY AND COMMUNICATION WITH THE PUBLIC

INFORMATION ABOUT YOUR WATER UTILITY

1 **Please provide the name of the person completing this survey and a phone number where we may reach you if we have any questions:**

Your Name: ____________________ Job Title: ____________________

Utility Name: ____________________ City: ____________________ State: ____________

Work Phone Number: (_____)____________________ Ext.: ____________

Email Address: ____________________ Web Site: ____________________

2 **Approximately how many <u>accounts</u> does your utility or system serve?** *(Please still enter a total even if you don't know the breakdown by type.)*

Households / residential	____________________
Commercial / industrial	____________________
Resale / wholesale to other utilities	____________________
Other *(please describe)* ______________	____________________
Total	____________________

3 **Which best describes the population area that your utility serves?** (*Check all ✓ that apply.*)

- ☐ Rural area
- ☐ Small city or town (not in a major metropolitan area)
- ☐ Suburban area (surrounding a major metropolitan area)
- ☐ Central city (downtown of a major metropolitan area)
- ☐ Other *(please describe)* ____________________

4 **Approximately how many <u>people</u> in total (of all ages) does your utility serve?**

__________________ people served, in total

5 **Approximately what percentage of the drinking water you provide comes from each of the following sources:** (*Check all ✓ that apply.*)

Purchased water	______	%
Surface water	______	%
Groundwater	______	%
Other *(please describe)* ______________________	______	%
Total	**100**	**%**

6 **How much have each of the following affected or impacted the operations of your water utility in the past three years?** *(Please mark ✓ one answer for each concern.)*

	No Impact at All	Minor Impact	Moderate Impact	Large Impact	Very Large Impact
Waterborne disease contamination requiring public notification	☐	☐	☐	☐	☐
Significant drought / water shortage . . .	☐	☐	☐	☐	☐
Source water protection issues	☐	☐	☐	☐	☐
Old or depleted infrastructure	☐	☐	☐	☐	☐
Regulatory requirements	☐	☐	☐	☐	☐
Water system security	☐	☐	☐	☐	☐
Financial or budget issues	☐	☐	☐	☐	☐
Other *(please describe)* ____________	☐	☐	☐	☐	☐

For any of these where you indicated a <u>Large Impact</u> or <u>Very Large Impact</u> please briefly describe what the concern or problem is and how this has affected or impacted your utility *(Please enclose extra pages if needed.)*

____

____

____

YOUR CONSUMER CONFIDENCE REPORT

7 **Who designs and prepares your CCR (e.g., who puts together your CCR not counting printing and mailing)?** *(Please mark ✓ one answer.)*

- ☐ We do this entirely "in-house" with our utility staff and resources.
- ☐ We hire an outside firm to assist our utility staff — we do more than ½ the work.
- ☐ We hire an outside firm to do most of the work — we do less than ½ the work.
- ☐ Other *(please describe)* ____________

8 **Who does the printing and mailing for your CCR?** *(Please mark ✓ one answer.)*

- ☐ We do this entirely "in-house" with our utility staff and resources.
- ☐ We hire an outside firm to help our utility staff — we do more than ½ the work.
- ☐ We hire an outside firm to do most of the work — we do less than ½ the work.
- ☐ Other *(please describe)* ____________

9 **On average, how many different versions of the CCR does your utility prepare each year (e.g., to different service areas)?**

______ different CCRs prepared each year.

10 **Approximately how much money is budgeted each year for preparing and distributing the CCRs?**

Preparation and design	$ ______
Printing	$ ______
Mailing costs	$ ______
Other costs (please describe)	$ ______
Total	$ ______

11 **Who do you send your CCR to?** (*Check ✓ all that apply.*)

- ☐ To all water utility accounts.
- ☐ To all mailing addresses in our service area.
- ☐ To all landlords and property owners in our service area.
- ☐ Other *(please describe)* ____________

12 **On average, how many total CCRs do you send out each year?**

______ CCRs sent out each year.

13 **How do you deliver your CCRs to the public?** (*Check ✓ all that apply.*)

- ☐ Separate first class mailing
- ☐ Separate third class mailing
- ☐ Separate bulk mailing
- ☐ Included with utility bill
- ☐ Other (*please describe*) ______________________________

14 **Besides actually sending the CCR out to your customers, which of the following methods does your utility use to publicize the CCR so that the public knows that it is available and is aware of the information in the CCR?** (*Check ✓ all that apply.*)

- ☐ Do not actively publicize our CCR
- ☐ Utility newsletter or bill stuffer
- ☐ Press releases to the local news media
- ☐ Public service announcements or ads on local television or radio
- ☐ Public service announcements or ads in local newspapers
- ☐ Community outreach activities and special events
- ☐ Other (*please describe*) ______________________________

____

15 **What, if anything, do you do that is different from most other utilities or that goes beyond the minimum requirements for the CCRs?** (*Please enclose extra pages if needed.*)

____

____

____

16 **How much impact do you think the CCR has had in helping improve customer perceptions of your water utility and the quality of the drinking water you provide?** (*Check ✓ your answer.*)

No Impact At All	Very Little Impact	Some Impact	Considerable Impact	Tremendous Impact	Don't Know
☐	☐	☐	☐	☐	☐

17 **Which of the following methods have you used to get input or feedback from customers or the general public regarding the CCR, either before or after it was prepared and distributed**? *(Check ✓ all that apply.)*

- ☐ Obtained feedback from customer advisory council or similar group
- ☐ Obtained feedback from public meetings or hearings on the topic
- ☐ Conducted focus groups and asked about the topic
- ☐ Conducted a survey and included questions on the topic
- ☐ Provided a phone number, mail address, or web site for customer questions or concerns
- ☐ Other *(please describe)* ________________________________
- ☐ None of the above

18 **In the past year, have you had any comments or questions from the public about your <u>*CCRs*</u>?** *(Check ✓ your answer.)*

- ☐ No → **Go to Question 19**
- ☐ Yes → if yes, how many comments did you receive about the following issues?

	No Comments or Questions	A Few Comments or Questions	Several Comments or Questions	A Lot of Comments or Questions
Too technical or difficult to understand . .	☐	☐	☐	☐
Question about spending money on CCRs	☐	☐	☐	☐
Like the information being provided	☐	☐	☐	☐
Concerns about water quality as reported in CCR	☐	☐	☐	☐
Did not receive a CCR	☐	☐	☐	☐
Requested additional copies of the CCR . .	☐	☐	☐	☐
Other *(please describe)* ______________	☐	☐	☐	☐

↳ **What important things did you learn from the public comments?** *(Please enclose extra pages if needed.)*

____

____

19 **In general, how much do you agree or disagree with the following comments about water quality and customer communications?** (*Check ✓ your answer for each concept.*)

	Disagree Completely	Disagree Somewhat	Neither Agree or Disagree	Agree Somewhat	Agree Completely	Doesn't Apply to My Utility
CCRs are a good way to inform customers about drinking water quality .	☐	☐	☐	☐	☐	☐
CCRs do more to scare customers than to inform them	☐	☐	☐	☐	☐	☐
Customers don't care about details of water quality as long as it is safe	☐	☐	☐	☐	☐	☐
The newly required arsenic language is important for informing customers about the risks of arsenic in water	☐	☐	☐	☐	☐	☐
We don't have sufficient resources to provide our customers all the information we would like them to have	☐	☐	☐	☐	☐	☐
The wording of the required arsenic language will only scare customers	☐	☐	☐	☐	☐	☐
Overall, providing CCRs to our residential customers has improved our relationship with them	☐	☐	☐	☐	☐	☐
CCRs are only a part of our overall responsibility to keep customers informed about water quality	☐	☐	☐	☐	☐	☐

ARSENIC AND YOUR CCR

20 **What was the maximum reported level of arsenic in your utility's drinking water for 2001?**

________ *µg/l (Please enter the level reported in your most recent CCR if available.)*

21 **If your 2001 testing showed an arsenic level between 5 and 10 µg/l, EPA requires you to provide an *<u>educational statement</u>* about arsenic in your CCR. If your 2001 testing showed an arsenic level above 10 µg/l and below 50 µg/l, you must include a *<u>health effects statement</u>* in your CCR. Does or will your 2001 CCR contain an *<u>educational or health effects statement</u>* about arsenic?** (*Check ✓ your answer.*)

☐ No statement required or included. → Please go to **23**

☐ The exact statement required by EPA. → Please go to **23**

☐ The exact statement required by EPA plus ***additional*** wording. → Please go to **22**

☐ An ***alternative*** to the statement required by EPA. → Please go to **22**

22 **If you included *additional* or *alternative wording* about arsenic in your CCR, was this language reviewed or approved by your primacy agency?** (*Check ✓ your answer.*)

☐ No → please go to **23**

☐ Yes → If yes, please briefly describe the process involved in obtaining approval *(please enclose extra pages if needed):*

____

____

____

____

23 **Whether or not special language was included in your CCR, have you received any comments from your customers regarding concerns about arsenic in their drinking water?**

☐ No comments at this time.

☐ Yes → If yes, how many comments did you receive about the following issues?

	No Comments or Questions	A Few Comments or Questions	Several Comments or Questions	A Lot of Comments or Questions
Want to know about current levels	☐	☐	☐	☐
Want information about health effects . . .	☐	☐	☐	☐
Other (________________)	☐	☐	☐	☐
Other (________________)	☐	☐	☐	☐

EMERGING ISSUES

24 **What issues do you see as being potential major concerns in the future of your water utility?** *(Please mark ✓ one answer for each concern.)*

	Definitely Not a Future Issue for My Water Utility	Possibly a Future Issue for My Water Utility	Definitely a **Minor** Future Issue for My Water Utility	Definitely a **Major** Future Issue for My Water Utility
Controlling waterborne diseases	☐	☐	☐	☐
Water supply shortages	☐	☐	☐	☐
Source water protection	☐	☐	☐	☐
Infrastructure replacement and rehabilitation	☐	☐	☐	☐
Changing regulatory requirements	☐	☐	☐	☐
Water system security	☐	☐	☐	☐
Providing water in unconventional or new manners (e.g., in house filter, bottled water)	☐	☐	☐	☐
Other *(please describe)* ______________	☐	☐	☐	☐

25 **Of the potential major concerns in the future of your water utility, which do you feel are most important and why?**

____

____

____

____

Thank you very much for your help with this survey. Please feel free to provide any additional comments or thoughts on additional sheets of paper.

Please return this questionnaire and <u>2 copies of your most recent Consumer Confidence Report</u> (if 2001 is not available, then please send your 2000 CCR) in the enclosed envelope to:

Marca Hagenstad
AwwaRF Consumer Confidence Report Project
Stratus Consulting Inc.
PO Box 4059
Boulder, CO 80306-4059

APPENDIX B
MICRO FOCUS GROUP DOCUMENTS

You and Your Household

Household information is used to help group your responses with responses of other households. Your individual responses and your name will not be released. *For each question, circle the number of your answer or fill in the blank space.*

1 **Your gender:** 1 Female 2 Male

2 **Your age:** ______ Years old

3 **What is the highest level of schooling you have completed?** *Circle the number of your answer.*

1 Did not complete high school
2 High school diploma or equivalent
3 Some college, two year college degree (AS) or technical school
4 Four year college graduate (BA, BS)
5 Some graduate work but did not receive a graduate degree
6 Graduate degree (MA, MS, MBA, PhD, JD, MD, etc.)

4 **What is you present employment status?** *Circle the number of your answer.*

1 Employed full time
2 Employed part time
3 Retired
4 Homemaker
5 Student
6 Unemployed

5 **During this session we will be talking about the quality of drinking water in your city. When you think about or discuss the "the quality of drinking water," what two or three things first come to mind?** *Please write your answers in the spaces below.*

____

____

____

6 **We're talking to water utility customers, at a number of locations around the country, trying to understand any concerns they might have about the quality of their water. Do you feel that the drinking water you receive from your water utility is safe to drink?** *(Please circle the number of your answer).*

1 Yes   If yes, please go to **7**

2 No   If no, please explain why not:

____

____

7 **Have you ever received any information from your water utility about the quality and /or safety of your drinking water?** *(Please circle the number of your answer).*

1 Yes   If yes, please tell us what information you received and when:

____

____

____

2 No, I have not received anything.

Focus Group Discussion Guide

I. Introduction

We are working for an association of water utilities. They want to help utilities around the country present technical (and other) information to customers in a way they can understand clearly. Tonight **today** we're going to talk about the quality of water here in **Boulder** ____________ and the kind of information you get from your utility about your water.

A. Purpose: Research – no sales

B. Confidentiality

C. Audio/video taping: client observation

D. Encourage participation — no right or wrong answers—no consensus—feel free to question one another

E. No side discussions

F. Questions?

II. Participant Introductions

A. First name

B. Occupation

C. Household size and make-up

D. What does "quality of drinking water" mean to you? – **"tell us an experience about a bad glass of water"**

III. Water Quality and Water Quality Information

We're talking to water utility customers, at a number of locations around the country, trying to get a good sense of any concerns they might have about the quality of their water.

In general, what do you think of the quality of your drinking water here in __________? [PROBE: How would you describe its Taste and Odor? What about Color & Clarity?

Do you have any concerns about the quality of your water? What concerns do you have about the quality of your water? [PROBE: Is it clean and safe?] How many of you drink bottled water? Why/Why not?

What kind of information do you receive from the utility about your water? Do you get enough information?

[Written exercise: Have participants respond in writing, then discuss] What would you like to know about your water? [Collect answers] How often would you like to receive information?

IV. Local Water Quality Report

A. How many of you remember getting an annual Water Quality Report from the utility in the past year? Did you look at it? What did it say? [PROBE: Did it say the quality of your water is good? Did it report any problems with water quality?]

PASS OUT WQR.

B. Take 30 seconds to look at this Water Quality Report. Does it look familiar to you? What information did you get from it?

C. Now take five **(seven-ten)** more minutes to read through it.

D. What is the main message you get from this report?

E. How does it make you feel about the quality of your water? [PROBE: Does it make you more concerned, less concerned, or not influence your concern either way?]

F. Pose questions based on written answers to question above about information needs. **Did it answer your questions?**

G. Is the Water Quality Report easy or difficult to understand? What makes it easy or difficult to understand? [PROBE: Is the language used too technical? Is the information organized in a logical way?] [Possible pre-testing of survey items.]

H. If you get something like this in the mail, are you more likely to read it or to just throw it away, unread? Why?

Scavenger Hunt: I'm giving each of you a piece of paper with some questions about the Water Quality Report on it. Please don't start answering until I give the word. The point of this exercise is to find out how easy it is for you to get information from the Water Quality Report.

1. Where does your water come from?
2. What does MCL refer to?
3. Are there any contaminants in your water that exceed maximum EPA limits?
4. **Are you safe from cryptosporidium contamination?**

5. What is the telephone number you could call to get more information about the quality of your local water?

I. Which of these pieces of information was difficult to find? Why?

J. What information is the most useful?

K. How does this information affect your thoughts and feelings about your water utility? What is it about the Water Quality Report that makes you feel that way?

L. What questions or concerns about the report itself do you have?

M. Can you suggest any ways to improve the Water Quality Report? [PROBE: What additional information would you like to receive? Were the tables easy to understand?]

N. What would be the best way to get this information to you? Would you be willing to log onto the web to receive water quality information? Has anyone here used the web to get information about water quality or other facts pertaining to drinking water?

V. Alternative Water Quality Report

A. Now that you've seen your own Water Quality report, I'd like you look at another Water Quality Report from a different city. I'd like you to do the same thing you did with the first one, read it for 30 seconds and give me your initial reactions. [Pause] What information did you get from it? Now read through it for two more minutes.

B. Let's try to answer some of the same questions we answered for the first report. [Repeat questions from IV C.]

C. **COMPARISON**: [Written answers to each initial question, then discuss.]

1. Which report is more useful? Why? [Probe for specific reasons: placement of information, salience of information, difficulty of language, page format.]
2. Which of the data tables do you prefer? Why?
3. Which report is visually more appealing? Why? [Probe for specific reasons: graphics, photos, use of color, layout, type fonts.]
4. Which report would you be more likely to pay attention to when it arrives in the mail? Why?
5. Which utility seems like a better-run utility? Why?
6. Overall, which Water Quality Report do you like better? **What would you tell the other utility they should do to improve their Water Quality Report?**

VI. Arsenic Section

In January of 2001 the Environmental Protection Agency issued a new rule that established a Maximum Contaminant Level for arsenic. The new level is 10 micrograms per liter (μg/L). The previous standard for arsenic was 50 μg/L. Systems that have arsenic levels above the new maximum are required to include a statement about the health effects of arsenic. Here's what the statement says. [SHOW STATEMENT AND READ.]

> Some people who drink water containing arsenic in excess of the MCL over many years could experience skin damage or problems with their circulatory system, and may have an increased risk of getting cancer.

Can you tell me in your own words what you think this statement is saying?

If your water did have arsenic levels above the new minimum, and you read this statement, what would you think and how would you feel? [PROBE: About your water? About your utility?]

Is there any action you would take?

Can you point me to any particular words or phrases that concern you?

The new rule also requires a statement in the Water Quality Report when a water supply meets the official standard of 10 μg/L, but is above 5 μg/L. Here's what the statement says. [SHOW STATEMENT AND READ.]

> While your drinking water meets EPA's standard for arsenic, it does contain low levels of arsenic. EPA's standard balances the current understanding of arsenic's possible health effects against the costs of removing arsenic from drinking water. EPA continues to research the health effects of low levels of arsenic, which is a mineral known to cause cancer in humans at high concentrations and is linked to other health effects such as skin damage and circulatory problems.

What would you think and how would you feel if you did read a statement such as this in your water quality report? [PROBE: About your water? About your utility?]

Is there any action you would take?

Can you point me to any particular words or phrases that concern you?

VII. Wrap-Up and Final Comments

Information About Contaminants in Drinking Water

In 2001 the Environmental Protection Agency issued a new rule that established a Maximum Contaminant Level (MCL) for arsenic. The new level is 10 micrograms per liter (μg/L). Systems that have arsenic levels <u>above</u> the new maximum are required to include a statement about the health effects of arsenic in their annual Water Quality Report. Here's what the statement says:

> **Some people who drink water containing arsenic in excess of the MCL over many years could experience skin damage or problems with their circulatory system, and may have an increased risk of getting cancer.**

1 **If your water had arsenic levels above the MCL, and you read this statement, what would you think and how would you feel about your water utility and about your drinking water?**

____

____

2 **What action, if any, you would take?**

____

____

3 **Are there any particular words or phrases in this statement that concern you or that you don't understand?**

____

____

The new EPA rule also requires a statement in the Water Quality Report when a water supply meets the official standard of 10 μg/L, but is above 5 μg/L. Here's what the statement says.

> **While your drinking water meets EPA's standard for arsenic, it does contain low levels of arsenic. EPA's standard balances the current understanding of arsenic's possible health effects against the costs of removing arsenic from drinking water. EPA continues to research the health effects of low levels of arsenic, which is a mineral known to cause cancer in humans at high concentrations and is linked to other health effects such as skin damage and circulatory problems.**

4 **If your water had arsenic levels below the MCL but above 5 μg/L, and you read this statement, what would you think and how would you feel about your water utility and about your drinking water?**

____

____

____

5 **What action, if any, you would take?**

____

____

____

6 **Are there any particular words or phrases in this statement that concern you or that you don't understand?**

____

____

____

APPENDIX C
ATTRIBUTE CHARACTERIZATION GUIDE

This appendix provides a complete list of CCR attributes identified for this study. The associated scoring method used for each attribute is also shown. The attribute numbers are not listed in a complete numerical sequence due to several revisions made to the attribute list over the course of this project.

1. Utility Company Characteristics

1.1 Name
1.6 Web Site Address
1.10 Contact Telephone
1.13 Type of Water Source (G = Ground Water Only; S = Surface Water only; GS = Both Ground and Surface Water)

2. Format/Functionality

2.2 Type of Document [Brochure = B, Web site = W]
2.2a Is front page eye-catching? (1 = not eye-catching; 2 = somewhat eye-catching; 3 = very eye-catching)
2.3 Document Length (# of Pages)
2.4 Font Size of Main Text (1 = < 12; 2 = 12-13; 3 = 14)
2.5 Font Size of Headings (1 = 12; 2 = 13-14; 3 = 15)
2.6 Delineation of Headings with Shaded Boxes, Shadow Effect or Other Methods? (Y/N)
2.7 Photos? (1 = No; 2 = Yes, one or two; 3 = Yes, more than two)
2.8 Graphics (other than maps)? (1 = No; 2 = Yes, one or two; 3 = Yes, more than two)
2.9 Maps? (Number of Maps)
2.10 Symbols? Y/N (e.g., are symbols used to help reader understand at a glance what a particular topic or heading means?)
2.11 Overall Style (1 = plain; 2 = moderate; 3 = glossy/fancy)
2.12 Text/Layout Style (1 = Textbook/Report; 2 = Newsletter (in columns); 3 = Brochure)
2.13 Color?
2.13a # of colors (1 = 1-2 colors; 2 = 3-6 colors; 3 = >6 colors)
2.14a Clarity of Language (1 = jargon/technical/statistical; 2 = moderately clear; 3 = very clear)
2.14b If not clear, why? (if 1 or 2 in 2.14a, explain why)
2.15a Logical flow or progression of the text and data (1 = confusing, 2 = moderately clear, 3 = clear) Note: please pull out example of confusing vs. clear flow of text and data.
2.15b If not clear, why? (if 1 or 2 in 2.14a, explain why)
- 2.16.1 Number of sentences (Give #)
- 2.16.2 Number of words with 3 or more syllables (Give #)
- 2.16.3 Gunning Fog Index (equals ((100 / question 2.16.1) + question 2.16.2) * 0.4

2.17 White Space (1 = high amount of white space; 3 = low amount of white space)

3. General Content

3.1 Date of Report (Year)
3.2 Date of Data in Report (Year)
3.3 Report Title (WQ = Water Quality Report; CCR = Consumer Confidence Report; B = Both, O = Other)
3.4 Contaminants
- 3.4.1 How many detected contaminants (if any) were in violation of the MCL, treatment technique, or exceeded an Action Level? (give #)
- 3.4.2 Gives names of detected contaminants over MCL

3.5 For Each Required Topic Area Below, except where noted, is the information: 1 = Not there; 2 = Somewhat visible (discussed in the text, but not easy to find); 3 = Very visible (listed in a heading or separate text box or table)

3.5.1 Water Supply Source(s)

3.5.2 Statement on Drinking Water Quality and Violations (e.g., heading might read "Is Your drinking water is safe to drink?" "Does your water meet EPA drinking water standards?" "Were there any violations of EPA drinking water standards?")

3.5.3a Health Information/Concerns/Effects on *Cryptosporidium*

3.5.3b Health Information/Concerns/Effects on radon

3.5.3c Health Information/Concerns/Effects on copper

3.5.3d Health Information/Concerns/Effects on lead

3.5.3e Health Information/Concerns/Effects on Trihalomethane

3.5.3f Health Information/Concerns/Effects on other contaminants (Give # of other contaminants discussed)

3.5.3g What are other contaminants discussed? (List names of other contaminants)

3.5.4 Health Information/Concerns/Effects for Arsenic.

3.5.4.2a Arsenic statements: health or educational

3.5.4.2b Arsenic statements: health or educational (N = none; H = health; E = educational; O = other)

3.5.4.3a Arsenic level in water quality data

3.5.4.3b What is the highest arsenic level detected? (list number and units or nd = non-detected)

3.5.5 Contact Information

3.5.6 Drinking Water Analysis Data — for Detected Contaminants in Treated Water

3.5.7 Drinking Water Analysis Data — for Detected Unregulated Substances

3.5.8 Definitions of Measurement Units and Terms

3.5.9 Information Presented in Different Languages

3.5.10 Information on regularly scheduled opportunities for public participation

3.5.11 Health/Educational Information on other contaminants (1st version only)

3.6 Content

3.6.1a Content Overview, Executive Summary, or Document Map

3.6.1b Table of contents

3.6.2 Description of the Drinking Water Standards/What they Mean/How they are Implemented

3.6.3 Frequency of Water Monitoring

3.6.4 Location of Water Sampling

3.6.5 Water Treatment Process Description

3.6.6 Other Measures Being Taken to Improve Drinking Water

3.6.8 List of Contaminants that Were Not Detected in Treated Water

3.6.9 Water System Maps (1st version only)

3.6.9a Map is more for Style than Information Content (Y/N)

3.6.9b Map is more for Informational Content (Y/N)

3.6.10 FAQ Section

3.6.11 General comparative information so reader can understand the context of water quality results for his location? Y/N

3.6.12 Description of Contaminants and their Sources
3.6.13 Protective Actions Consumers Can Take to Ensure Safe Water
3.6.14 Variations in Contaminant Level Over Time
3.6.15 Description of Water System Security Issues
3.6.16 Point-of-use (POU), home filter, and other home treatment devices.
3.6.17 Other supplemental information (Y/N; provide description of other information)
3.6.18 What is other supplemental information?

4. Specific Content

4.1 Tone of Health Effects Discussion (1 = pessimistic; 2 = optimistic; 3 = realistic)
4.2 Contact Information
4.2.1 Location of contact information (F = first page; L = last page; FL = both; O = other)
Does it include:
4.2.2 Telephone numbers (Y/N)
4.2.3 E-mail addresses (Y/N)
4.2.5 Separate opportunity to provide comments (e.g., postcard, survey form) (Y/N)
4.2.6 24-hour emergency telephone number (Y/N)
4.3 Water Quality Table(s) Data
4.3.1 Presented as aggregate or zone? (A = aggregate; Z = zone)
4.3.3 Dates Data Were Collected? (Y/N)
4.3.4 Is it clear whether or not violations have occurred for each detected contaminant? (1 = not there; 2 = somewhat visible/there but hard to find; 3 = very visible)
4.3.5 Are symbols or graphics used to help reader see if there were any violations or other concerns related to water quality? (Y/N)
4.3.6 Are data presented as whole numbers (W) or as decimals (D)?
4.3.7 Is there a description or other explanatory text about how to read the data table? (Y/N)
4.4 Information Presented in Other Languages
4.4.1a How many different languages? (Give number)
4.4.1b Spanish? (Y/N)
4.4.1c French? (Y/N)
4.4.1d Other? (specify languages presented)
4.4.2 Type of Information:
(4.4.2a) Topic of the report (i.e., water quality in your community) (Y/N)
(4.4.2b) Contact information (Y/N)
(4.4.2c) Other information (Y/N)
4.5.2 Water System Maps
If provided, does map show location of:
(4.5.2a) Individual Wells (g = general location; s = specific location; n = no)
(4.5.2b) Surface Water Sources (e.g., reservoirs, streams, rivers) (g = general location; s = specific location; n = no)
(4.5.2a) Does map show locations of pipelines and/or reservoirs? (Y/N)
(4.5.2a) Does map show locations of treatment facilities? (Y/N)

APPENDIX D
METHODOLOGY FOR THE NATIONAL PHONE SURVEY

This appendix contains more detailed information on the survey sample and methodology used for the national phone survey (presented in Chapter 6) conducted as part of this AwwaRF project to measure awareness and perceptions of CCRs and to assess their effectiveness and impact on consumer attitudes toward and perceptions of drinking water and local water utilities. Included here are descriptions of (1) survey goals and objectives; (2) survey design, questions, and topics; (3) survey methods and sampling; (4) sample composition and weighting; and (5) survey analysis and reporting.

SURVEY GOALS AND OBJECTIVES

The primary objectives for the national phone survey were to

- Use a statistically valid, random, and representative sample of U.S. adults nationwide
- Determine how many people recalled getting a CCR
- Measure their reactions to and perceptions of that report
- Assess the impact of CCRs on customer perceptions of water quality, information, and local water utilities by:
  - Comparing these perceptions among customers who do and do not recall getting a CCR
  - Tracking changes in these perceptions since the CCR requirements were implemented in 1999

Other key goals for the project overall included identifying the characteristics, attributes, or conditions that make for more successful CCRs. Although this was best addressed by the central site survey in Chapter 7, we were also able to generate some relevant information and analysis as part of the national phone survey. This information was analyzed as part of an effort to identify the "attributes" or characteristics that make CCRs more usable for consumers and more effective for water utilities (see Chapter 5 for more discussion of these concepts and distinctions).

Note that the project was expanded to address issues related to consumer reaction to health effects language about arsenic. In addition, the project explored consumer use and perceptions of POU devices and their interest in related products and services that local water utilities might offer.

SURVEY DESIGN, QUESTIONS, AND TOPICS

The survey was directed by Robert Hurd (Hurd & Associates, Inc.) and designed to ensure that the survey sample, methods, and results were comparable to other national phone surveys he conducted for AwwaRF in 1993 and 1997-1998. The questionnaire repeated key tracking questions from these previous surveys for comparison purposes and to show changes in consumer attitudes and perceptions since CCRs were first required nationwide in 1999.

The survey was developed in consultation with Jeffrey Lazo (Stratus Consulting), the principal investigator, and other members of the research team. Survey development included an effort to incorporate preliminary results from the literature review (Chapter 2), the water utility survey (Chapter 3), the customer focus groups (Chapter 4) and the analysis of CCR content and attributes (Chapter 5). The survey questionnaire was then reviewed and approved by the AwwaRF project manager and PAC.

To identify and correct any possible problems with the survey questions, sampling, or interviewing the project team conducted a formal "pre-test" of the survey with a random sample of roughly 50 respondents before full-scale interviewing began. No major problems were identified and the results were included in the final sample total.

Survey topics included customer satisfaction, perceptions of water quality and safety, and awareness and perceptions of the information people received from their local water utilities, including the annual water quality reports (formally called Consumer Confidence Reports in the 1998 SDWA Amendments that require their preparation and distribution by local water utilities).

As noted previously, the survey was also expanded to include additional questions to assess customer reactions to the wording of health effects language for arsenic, and explore in more detail issues surrounding POU and filter devices.

SURVEY METHODS AND SAMPLING

The survey was administered by telephone to a random sample of 1,414 adult U.S. residents nationwide during December 2002. The total weighted sample of 1,414 respondents included 1,146 "customers" who get their tap water from a local water utility, plus 268 individuals who rely on their own private well.

Experienced, trained professionals from Woelfel Research, Inc., and its fieldwork centers in northern Virginia and central Pennsylvania conducted the telephone interviews. The interviews averaged between 8 and 15 minutes in length, depending on the type of respondent and the number of survey questions that applied to each group.

To cover a wider range of topics in greater depth, the research team developed two different versions of the survey and administered them to two independent random samples of roughly 700 respondents each ["phase one" (P1) and "phase two" (P2)]. Each phase included about 550 customers who get their tap water from a local water utility—and about 150 people who rely on their own private well.

Results from the two samples were combined for analysis purposes, creating a single random sample of 1,414 adults nationwide. This sample was representative of the total population and relevant subgroups (including water utility customers, private well users, and subgroups based on demographics or other characteristics).

Some additional efforts were made to identify and interview a somewhat larger number of water utility customers (and fewer households that rely on private wells) than would otherwise be obtained in a random sample of adults nationwide. This helped in our analysis by increasing the sample size for this important subgroup, thereby enhancing the accuracy or level of confidence in those results. This oversampling of customers had some impact on the initial demographics of the sample, but this was largely corrected through statistical weighting in the final total sample (see the discussion that follows).

SAMPLE COMPOSITION AND WEIGHTING

The demographics of the survey sample compared favorably with known population parameters based on U.S. Census projections and previous national surveys conducted for AwwaRF (Hurd 1993, 1997, 1999, 2003). As with the previous surveys in 1993 and 1997-1998, however, the survey sample was also statistically weighted using standard industry procedures to correct for minor deviations in certain demographic characteristics (gender, age, and geographic region).

Table D.1
Sample size and margin of error for key subgroups

Size of sample or subgroup	Subgroups included	Margin of error (%)
1,400 to 1,450	All adults in total sample	± 2.6
1,100 to 1,150	All customers in total sample	± 2.9
700 to 710	All adults in P1 or P2 sample	± 3.7
550 to 570	All customers in P1 or P2 sample	± 4.1
430 to 440	All customers who recall a CCR	± 4.7
260 to 300	All well users in total sample	± 6.0

These adjustments corrected for the oversampling of customers and for differential response among demographic subgroups. Weighting the final sample helped to ensure that the sample and results were comparable to previous AwwaRF surveys, and more accurately reflected the proportions found in the total population of adult residents and water utility customers nationwide.

SAMPLE SIZE AND MARGIN OF ERROR

Survey results from the total sample of 1,414 adults nationwide are accurate within ± 2.6 percentage points—the margin of sampling error allowed for a sample of this size (n = 1,414) at the 95% confidence level. Results for subgroups have somewhat larger margins of error, depending on the size of each group or subsample (see Table D.1).

Note that different groups of respondents were sometimes asked different survey questions. For that reason, the sample size or base for some survey results may vary, depending on the total number of respondents who were asked a particular question.

ANALYSIS AND REPORTING

The national phone survey includes a mix of survey questions that measure respondent characteristics and perceptions at the ordinal and interval levels of measurement. Most results are presented as percentages—sometimes as a percentage of a certain subgroup asked a particular question (i.e., people who recall getting a CCR), and sometimes as a percentage of the total sample (either all adults or all customers).

Note that the total sample or base for some questions and subgroups in the weighted sample may vary slightly, and some results may not always add to 100% because of rounding.

The mean (or average) is also provided (along with the scale) when such statistics are relevant or appropriate for the type of question asked. The "not sure" option is not included when calculating these statistics.

Most of the statistical analysis involved making comparisons between subgroups of customers to test for significant differences in their perceptions or responses. For example, we compared men and women (and other subgroups based on demographics), as well as customers who did and did not recall getting a CCR.

For convenience and ease of presentation, these comparisons were made using either ANOVA techniques (for comparing more than two groups), or "t-test" procedures (for comparing two groups or items). These procedures determine whether there were statistically significant differences between groups (or items) in their mean or average level of response to some survey question or scale. Other procedures used in this analysis include correlation and regression to assess the strength of relationships among responses to survey questions or measures.

These procedures tend to be "robust" to assumptions of measurement at the interval level. In some cases where statistical procedures were used despite these assumptions, other (nonparametric) procedures were used to confirm and validate results. Few, if any, differences were found based on the different types of methods and assumptions.

APPENDIX E
CENTRAL SITE SURVEY

WATER QUALITY AND CUSTOMER COMMUNICATIONS

To help utilities provide better information to their customers, the Awwa Research Foundation has sponsored this study of water quality reports. You will be asked to read and evaluate two water quality reports from two different U.S. cities. Please read the reports carefully and then answer the questions on this response form. Before reading the first report, please answer the following 10 questions about your current satisfaction with your own water service.

Current Satisfaction

1 **In the past 12 months, have you experienced problems with any of the following at your home?** *(Circle one number for each)*

The water pressure at your tap	1 Yes	2 No
The taste of your tap water	1 Yes	2 No
The odor of your tap water	1 Yes	2 No
The color, clarity, or appearance of your tap water	1 Yes	2 No
The health and safety of your tap water	1 Yes	2 No

2 **How would you rate the quality of the drinking water in your community?** *(Circle one number)*

1 Excellent
2 Good
3 Fair
4 Poor
5 Not sure

3 **Overall, how satisfied are you as a customer with your local water utility?** *(Circle one number)*

1 Totally satisfied
2 Mostly satisfied
3 Somewhat satisfied
4 Not very satisfied
5 Not at all satisfied
6 Not sure

4 **Overall, how much trust do you have in your local water utility? Do you trust it:** *(Circle one number)*

1 Totally
2 Mostly
3 Somewhat
4 Not very much
5 Not at all
6 Not sure

5 **How often do you receive information from your water utility about the quality of your drinking water?** *(Circle one number)*

1 Once every month
2 Once every 6 months
3 Once every year
4 Less often than once every year
5 Never
6 Other (please specify________________________________)
7 Don't know

6 **How much information about the quality and safety of your drinking water do you get from your water utility?** *(Circle one number)*

1 A lot of information
2 Some information
3 Very little information
4 No information at all
5 Not sure

7 **Overall, how satisfied are you with the information you get from your water utility?** *(Circle one number)*

1 Totally satisfied
2 Mostly satisfied
3 Somewhat satisfied
4 Not very satisfied
5 Not at all satisfied
6 Not sure

Most water utilities are required by the U.S. Environmental Protection Agency (or EPA) to send a written report to their customers each year on the quality and safety of the local tap water they provide. Sometimes this water quality report is included with the customer's water bill, and sometimes it is mailed separately to their home.

8 **Can you remember if your household has ever received a water quality report from your water utility?** *(Circle one number)*

1 Yes
2 No
3 Not sure

9 **Did your household get a water quality report from your water utility sometime in the past year?** *(Circle one number)*

1 Yes
2 No
3 Not sure

10 **Does your household directly pay the water utility for tap water service or is that bill paid by someone else, such as a landlord?** *(Circle one number)*

1 Household directly pays water bill
2 Someone else pays water bill
3 Not sure

Stop!

Please Wait For Further Instructions

FIRST WATER QUALITY REPORT

In the envelope, you will find two Water Quality Reports that are used by two U.S. cities to communicate with its customers about the quality and safety of their local tap water. Please take either one of the reports out of the envelope, leaving the other one in the envelope for now. We would like you to spend 10 minutes reading this report. Then we would like your feedback about it.

11 PLEASE WRITE THE IDENTIFICATION (ID#) NUMBER LOCATED ON THE UPPER RIGHT-HAND CORNER OF THE WATER QUALITY REPORT IN THIS BOX ➔ ID#

Based on your reading of the Water Quality Report, please answer the following questions.

12 **How useful is the information provided in this report?** *(Circle one number)*

Not at all useful										Completely useful
0%	10%	20%	30%	40%	50%	60%	70%	80%	90%	100%

13 **How well organized is the information within this report?** *(Circle one number)*

Not at all well organized										Completely well organized
0%	10%	20%	30%	40%	50%	60%	70%	80%	90%	100%

14 **How easy to read and understand is the language in this report?** *(Circle one number)*

Not at all understandable										Completely understandable
0%	10%	20%	30%	40%	50%	60%	70%	80%	90%	100%

15 **How attractive is the appearance of this report?** *(Circle one number)*

Not at all attractive										Completely attractive
0%	10%	20%	30%	40%	50%	60%	70%	80%	90%	100%

16 **Overall, how satisfied are you with the report and the way it presents information to you?** *(Circle one number)*

Not at all satisfied										Completely satisfied
0%	10%	20%	30%	40%	50%	60%	70%	80%	90%	100%

17 **If this were your own city, how would you feel about the quality and safety of your water after reading this report?** *(Circle one number)*

1 A lot more confident
2 A little more confident
3 Neither more nor less confident
4 A little less confident
5 A lot less confident

Next, thinking about the same water quality report you've just read, we want to ask you about some phrases that might or might not be used to describe this water quality report. For each phrase, please tell us how much you agree or disagree that it describes this water quality report. *(Circle one number)*

This water quality report . . .

18 **Would stand out and get your attention in the mail.** *(Circle one number)*

Completely agree	Mostly agree	Somewhat agree	Somewhat disagree	Mostly disagree	Completely disagree	Not sure
1	2	3	4	5	6	9

19 **Is a waste of money.** *(Circle one number)*

Completely agree	Mostly agree	Somewhat agree	Somewhat disagree	Mostly disagree	Completely disagree	Not sure
1	2	3	4	5	6	9

20 **Is something you would want to keep for future reference.** *(Circle one number)*

Completely agree	Mostly agree	Somewhat agree	Somewhat disagree	Mostly disagree	Completely disagree	Not sure
1	2	3	4	5	6	9

21 **Is professional.** *(Circle one number)*

Completely agree	Mostly agree	Somewhat agree	Somewhat disagree	Mostly disagree	Completely disagree	Not sure
1	2	3	4	5	6	9

This water quality report . . .

22 **Could be mistaken for junk mail.** *(Circle one number)*

Completely agree	Mostly agree	Somewhat agree	Somewhat disagree	Mostly disagree	Completely disagree	Not sure
1	2	3	4	5	6	9

23 **Is credible and believable.** *(Circle one number)*

Completely agree	Mostly agree	Somewhat agree	Somewhat disagree	Mostly disagree	Completely disagree	Not sure
1	2	3	4	5	6	9

24 **Leaves questions unanswered or gives only part of the story.** *(Circle one number)*

Completely agree	Mostly agree	Somewhat agree	Somewhat disagree	Mostly disagree	Completely disagree	Not sure
1	2	3	4	5	6	9

25 **Is designed more for public relations than to provide customers with real information.** *(Circle one number)*

Completely agree	Mostly agree	Somewhat agree	Somewhat disagree	Mostly disagree	Completely disagree	Not sure
1	2	3	4	5	6	9

26 **Is easy to read, clear, and understandable.** *(Circle one number)*

Completely agree	Mostly agree	Somewhat agree	Somewhat disagree	Mostly disagree	Completely disagree	Not sure
1	2	3	4	5	6	9

This water quality report . . .

27 **Was produced only because the utility was required to produce it.** *(Circle one number)*

Completely agree	Mostly agree	Somewhat agree	Somewhat disagree	Mostly disagree	Completely disagree	Not sure
1	2	3	4	5	6	9

28 **Is too technical for you to understand.** *(Circle one number)*

Completely agree	Mostly agree	Somewhat agree	Somewhat disagree	Mostly disagree	Completely disagree	Not sure
1	2	3	4	5	6	9

29 **Gives enough information to decide whether or not the water is safe to drink.** *(Circle one number)*

Completely agree	Mostly agree	Somewhat agree	Somewhat disagree	Mostly disagree	Completely disagree	Not sure
1	2	3	4	5	6	9

Stop!
Please Wait For Further Instructions

SECOND WATER QUALITY REPORT

In the envelope, please take out the second Water Quality Report that is used by another U.S. city to communicate with its customers about the quality of their water. Please put the first report back in the envelope. We would like you to spend 10 minutes reading this second report. Then we would like your feedback about it.

30 PLEASE WRITE THE IDENTIFICATION (ID#) NUMBER LOCATED ON THE UPPER RIGHT-HAND CORNER OF THE WATER QUALITY REPORT IN THIS BOX → **ID#** ______

Based on your reading of the Water Quality Report, please answer the following questions.

31 **How useful is the information provided in this report?** *(Circle one number)*

Not at all useful										Completely useful
0%	10%	20%	30%	40%	50%	60%	70%	80%	90%	100%

32 **How well organized is the information within this report?** *(Circle one number)*

Not at all well organized										Completely well organized
0%	10%	20%	30%	40%	50%	60%	70%	80%	90%	100%

33 **How understandable is the language in this report?** *(Circle one number)*

Not at all understandable										Completely understandable
0%	10%	20%	30%	40%	50%	60%	70%	80%	90%	100%

34 **How attractive is the appearance of this report?** *(Circle one number)*

Not at all attractive										Completely attractive
0%	10%	20%	30%	40%	50%	60%	70%	80%	90%	100%

35 **Overall, how satisfied are you with the report and the way it presents information to you?** *(Circle one number)*

Not at all satisfied										Completely satisfied
0%	10%	20%	30%	40%	50%	60%	70%	80%	90%	100%

36 **If this were your own city, how would you feel about the quality and safety of your water after reading this report?** *(Circle one number)*

1 A lot more confident
2 A little more confident
3 Neither more nor less confident
4 A little less confident
5 A lot less confident

Next, thinking about the same water quality report you've just read, we want to ask you about some phrases that might or might not be used to describe this water quality report. For each phrase, please tell us how much you agree or disagree that it describes this water quality report. *(Circle one number)*

This water quality report . . .

37 **Would stand out and get your attention in the mail.** *(Circle one number)*

Completely agree	Mostly agree	Somewhat agree	Somewhat disagree	Mostly disagree	Completely disagree	Not sure
1	2	3	4	5	6	9

38 **Is a waste of money.** *(Circle one number)*

Completely agree	Mostly agree	Somewhat agree	Somewhat disagree	Mostly disagree	Completely disagree	Not sure
1	2	3	4	5	6	9

39 **Is something you would want to keep for future reference.** *(Circle one number)*

Completely agree	Mostly agree	Somewhat agree	Somewhat disagree	Mostly disagree	Completely disagree	Not sure
1	2	3	4	5	6	9

40 **Is professional.** *(Circle one number)*

Completely agree	Mostly agree	Somewhat agree	Somewhat disagree	Mostly disagree	Completely disagree	Not sure
1	2	3	4	5	6	9

This water quality report . . .

41 **Could be mistaken for junk mail.** *(Circle one number)*

Completely agree	Mostly agree	Somewhat agree	Somewhat disagree	Mostly disagree	Completely disagree	Not sure
1	2	3	4	5	6	9

42 **Is credible and believable.** *(Circle one number)*

Completely agree	Mostly agree	Somewhat agree	Somewhat disagree	Mostly disagree	Completely disagree	Not sure
1	2	3	4	5	6	9

43 **Leaves questions unanswered or gives only part of the story.** *(Circle one number)*

Completely agree	Mostly agree	Somewhat agree	Somewhat disagree	Mostly disagree	Completely disagree	Not sure
1	2	3	4	5	6	9

44 **Is designed more for public relations than to provide customers with real information.** *(Circle one number)*

Completely agree	Mostly agree	Somewhat agree	Somewhat disagree	Mostly disagree	Completely disagree	Not sure
1	2	3	4	5	6	9

45 **Is easy to read, clear, and understandable.** *(Circle one number)*

Completely agree	Mostly agree	Somewhat agree	Somewhat disagree	Mostly disagree	Completely disagree	Not sure
1	2	3	4	5	6	9

This water quality report . . .

46 **Was produced only because the utility was required to produce it.** *(Circle one number)*

Completely agree	Mostly agree	Somewhat agree	Somewhat disagree	Mostly disagree	Completely disagree	Not sure
1	2	3	4	5	6	9

47 **Is too technical for you to understand.** *(Circle one number)*

Completely agree	Mostly agree	Somewhat agree	Somewhat disagree	Mostly disagree	Completely disagree	Not sure
1	2	3	4	5	6	9

48 **Gives enough information to decide whether or not the water is safe to drink.** *(Circle one number)*

Completely agree	Mostly agree	Somewhat agree	Somewhat disagree	Mostly disagree	Completely disagree	Not sure
1	2	3	4	5	6	9

Please put the second water quality report back into the envelope.

Stop!
Please Wait For Further Instructions

About You

This final series of questions asks about you and your household's characteristics. The information is used for classification purposes. Your answers will be kept confidential and combined with those of others to get a general picture of our respondents.

49 **Are you male or female?** *(Circle one answer)*

Male Female

50 **In what year were you born?** *(Write the year on the line at right)*________________

51 **What is the highest level of education you have completed?** *(Circle one number)*

1 Did not complete high school
2 Completed high school (or GED)
3 Some college or technical school, no degree
4 Two year college degree or technical school degree
5 Four year college degree
6 Masters or doctoral work or degree

52 **Are there children under 18 years of age living in your home?** *(Circle one answer)*

Yes No

53 **Are there children under 3 years of age living in your home?** *(Circle one answer)*

Yes No

54 **Which of the following best describes your racial or ethnic identity?** *(Circle all that apply)*

1 White
2 African American or Black
3 Hispanic
4 Asian
5 Native American
6 Other________________

55 **Which of the following best describes your total 2002 household income before taxes and other deductions?** *(Circle one number)*

1 Under $20,000
2 $20,000 to $39,999
3 $40,000 to $69,999
4 $70,000 or more

Stop!
Please Wait For Further Instructions

APPENDIX F
COMPLETE ARSENIC REPORT

Arsenic Rule: Consumer Confidence Report Guidance

Arsenic Rule: Consumer Confidence Report Guidance

Prepared by:
Jeffrey K. Lazo, **Marca L. Hagenstad**, **Charles N. Herrick**, **Joanna L. Pratt**, and **Robert S. Raucher**
Stratus Consulting Inc.
P.O. Box 4059, Boulder, CO 80306-4059

Sponsored by:
Awwa Research Foundation
6666 West Quincy Avenue, Denver, CO 80235-3098

Published by:
Awwa Research Foundation
Advancing the Science of Water®

DISCLAIMER

CONTENTS

TABLES

FOREWORD

The Awwa Research Foundation is a nonprofit corporation that is dedicated to the implementation of a research effort to help utilities respond to regulatory requirements and traditional high-priority concerns of the industry. The research agenda is developed through a process of consultation with subscribers and drinking water professionals. Under the umbrella of a Strategic Research Plan, the Research Advisory Council prioritizes the suggested projects based upon current and future needs, applicability, and past work; the recommendations are forwarded to the Board of Trustees for final selection. The foundation also sponsors research projects through the unsolicited proposal process; the Collaborative Research, Research Applications, and Tailored collaboration programs; and various joint research efforts with organizations such as the U.S. Environmental Protection Agency, the U.S. Bureau of Reclamation, and the Association of California Water Agencies.

This publication is a result of one of these sponsored studies, and it is hoped that its finding will be applied in communities throughout the world. The following report serves not only as a means of communicating the results of the water industry's centralized research program but also as a tool to enlist the further support of the nonmember utilities and individuals.

Projects are managed closely from their inception to the final report by the foundation's staff and large cadre of volunteers who willingly contribute their time and expertise. The foundation serves a planning and management function, and awards contracts to other institutions such as water utilities, universities, and engineering firms. The funding for this research effort comes primarily from the Subscription Program, through which water utilities subscribe to the research program and make an annual payment proportionate to the volume of water they deliver and consultants and manufacturers subscribe based on their annual billings. The program offers a cost-effective and fair method for funding research in the public interest.

A broad spectrum of water supply issues is addressed by the foundation's research agenda: resources, treatment and operations, distribution and storage, water quality and analysis, toxicology, economics, and management. The ultimate purpose of the coordinated effort is to assist water suppliers to provide the highest possible quality of water economically and reliably. The true benefits are realized when the results are implemented at the utility level. The foundation's trustees are pleased to offer this publication as a contribution toward that end.

Federal regulations require all community water systems to deliver a CCR to their customers by July 1 of each year. In the Safe Drinking Water Act Amendments of 1996, Congress required USEPA to establish regulations that provide "brief statements in plain language regarding the health concerns that resulted in the regulation of each regulated contaminant." Congress required CCRs to include this language "for any regulated contaminant for which there has been a violation" of the Maximum Contaminant Level (MCL). In January 2001, the USEPA lowered the MCL for arsenic to 10 μg/L from the previous standard for arsenic of 50 μg/L. USEPA's new arsenic rule also created new requirements for public notification and language requirements for inclusion in CCRs. These communication requirements became effective February 22, 2002.

The purpose of this project is to provide guidance to community water systems on how to address Consumer Confidence Report (CCR) and risk communication requirements related to arsenic. This work was conducted as an expansion of work on AwwaRF Project 2692: "*The Impact of CCRs and Emerging Communication Issues on Customers*." Work related specifically to arsenic was used to further evaluate and refine tools for effective communication specifically with respect to language on arsenic contamination required for inclusion in Consumer Confidence Reports.

Edmund G. Archuleta, P.E.	James F. Manwaring, P.E.
Chair, Board of Trustees	Executive Director
Awwa Research Foundation	Awwa Research Foundation

ACKNOWLEDGMENTS

This report was funded by the Awwa Research Foundation. The project was managed and administered by India Williams. The researchers would like to extend their great appreciation to AwwaRF, and especially to India Williams for her management, patience, and encouragement. We also wish to thank the project advisory committee (PAC), which provided very useful comments, guidance, and critiques of this project. The PAC furnished comments on previous reports on this project and generally helped shape and improve this effort. The PAC members were:

Kelly Enders of Medical Education Collaborative

Randolph (Randy) Harmon of Birmingham Water Works

Christine Hoover of The Office of Consumer Advocate, Office of Attorney General, Pennsylvania.

The researchers also wish to acknowledge the useful input and support provided by several individuals who helped review several portions of this work at various stages of its development, including Dr. Robert Hurd of Hurd and Associates, and Drs. Eric Rambo and Robert Baumgartner of PA Consulting. In addition, we would like to thank the participating utilities from AwwaRF Project 2962: Laurie Lentz of Austin, Texas; Ed Grusheski of Philadelphia, Pennsylvania; Joe Harbour of Bellevue, Washington; Khis Suravallop of Tacoma, Washington; Sue Atherton of Illinois-American (Peoria, Illinois); Kathy Allen of Cincinnati, Ohio; Larry Bingaman of Bridgeport, Connecticut — Aquarion Water Company of Connecticut; and Amy Struthers of Boulder, Colorado.

Last, but not least, the researchers wish to thank the many individuals at Stratus Consulting who helped execute the technical tasks and produce this report.

Thank you all very much.

Jeffrey K. Lazo, PhD (Principal Investigator)
Stratus Consulting Inc.
Boulder, Colorado (date)

EXECUTIVE SUMMARY

OBJECTIVES

The purpose of this project is to provide guidance to community water systems on how to address Consumer Confidence Report (CCR) and risk communication requirements related to the final arsenic rulemaking, issued by the U.S. Environmental Protection Agency (USEPA). This work was conducted as an expansion of work on AwwaRF Project 2692: "*The Impact of CCRs and Emerging Communication Issues on Customers*." Work related specifically to arsenic was used to further evaluate and refine tools for effective communication specifically with respect to language on arsenic contamination required for inclusion in Consumer Confidence Reports.

CONSUMER CONFIDENCE REPORTING REQUIREMENTS

Federal regulations require all community water systems to deliver a CCR to their customers by July 1 of each year. In the Safe Drinking Water Act Amendments of 1996, Congress required USEPA to establish regulations that provide "brief statements in plain language regarding the health concerns that resulted in the regulation of each regulated contaminant." Congress required CCRs to include this language "for any regulated contaminant for which there has been a violation" of the Maximum Contaminant Level (MCL).

The public's health is protected by limiting the levels of specific contaminants in water by setting an enforceable MCL for each regulated contaminant. The MCL is the maximum permissible level of a contaminant in water delivered to any user of a public water system. In January 2001, the USEPA lowered the MCL for arsenic to 10 μg/L from the previous standard of 50 μg/L [*Federal Register* 2001, 66 (14) 6976]. USEPA's new arsenic rule also created new requirements for public notification and language requirements for inclusion in CCRs. These communication requirements became effective February 22, 2002.

There are two different USEPA-mandated health effects statements for arsenic: a health effects statement and an educational statement. Required use of these statements in CCRs depends on the level of detected (not average) arsenic concentrations in the water. The required arsenic health effects language when arsenic is detected at or above 10 μg/L is:

> Some people who drink water containing arsenic in excess of the MCL over many years could experience skin damage or problems with their circulatory system, and may have an increased risk of getting cancer.

Congress also allowed USEPA to require CCRs to include a health effects statement even when systems were not violating an MCL. The USEPA mandated-educational statement for arsenic when arsenic is detected at or above 5 µg/L and below 10 µg/L is:

> While your drinking water meets EPA's standard for arsenic, it does contain low levels of arsenic. EPA's standard balances the current understanding of arsenic's possible health effects against the costs of removing arsenic from drinking water. EPA continues to research the health effects of low levels of arsenic which is a mineral known to cause cancer in humans at high concentrations and is linked to other health effects such as skin damage and circulatory problems.

ARSENIC HEALTH EFFECTS AND EDUCATIONAL STATEMENT RESEARCH

Research for this project comprised three tasks: (1) an informal utility survey, (2) focus groups with water utility customers, and (3) a national telephone survey of the general public.

The informal survey questionnaire was mailed to 401 drinking water utilities in June 2002 to elicit input regarding their concerns with the 2001 arsenic rule requirements for CCR language and to explore approaches used by different utilities in meeting these requirements; 118 completed responses were received. Many utilities indicated that they consider the required arsenic language important for informing customers about the risks of arsenic in water, while they also believe that the required arsenic language will only scare customers. A small portion of the utilities (4%) reported arsenic levels over 25 µg/L and thus were required to include the educational statement before the arsenic rule revisions. With the new arsenic rule, 9.3% of the utilities would need to include the educational statement because their arsenic levels are greater than 5 μg/L. Of the utilities currently including arsenic text in their 2001 CCR, most included the exact statement required by USEPA, three used an alternative statement, and one provided additional wording in addition to the required statement directing customers to a website where further information could be found. Roughly 10% of the utilities reported receiving customer comments concerning arsenic. Most of these customers wanted to know about current arsenic levels, while a few wanted to know about the health effects of arsenic.

Eleven group interviews, or "focus groups," were conducted with residential utility customers in five cities to elicit evaluative commentary from customers as they read and reacted to CCRs. Participants reviewed and commented on the required arsenic language. Most participants stated the arsenic health effects statement for greater than the MCL would cause them concern. Several commented the language would leave them confused and cause them to worry, reduce consumption of their tap water, and lose trust in their utility. Some individuals were especially concerned about the language's uncertainty over the level of risk. Particular phrases that were unclear include "some people," "in excess," and "circulatory problems." Participants also wanted more information on how they could safely use the water. Several focus group participants were concerned about their water after reading the educational statement for the 5 to 10 μg/L levels, but were a little less concerned than they were after reading the wording for the 10+ μg/L level. However, they were clearly disturbed that cost was included in the statement as an issue affecting their safety.

The telephone survey of 1,414 U.S. adult residents nationwide was designed to assess how water utility communications have affected public awareness and perceptions of drinking water quality and safety. A section of the survey focused primarily on issues related to health effects language with emphasis on the USEPA arsenic language and alternatives to the mandated arsenic language. Individuals were read two of seven explanations on the health effects of various potential contaminants in drinking water. Several explanations correspond exactly with USEPA health effects language, and some are modifications of the required language. Respondents were asked their degree of agreement or disagreement with five statements regarding the health effects explanations on how individuals perceived the importance, clarity, and content of the information in the explanations; how much worry the statement caused; and how likely the explanation was to cause them to change their drinking water habits.

Responses to the phone survey indicate that while a majority or respondents agreed that the information is "important and worthwhile" and "clear and understandable," they are somewhat less likely to agree that the statements provide "enough information to decide whether or not your water is safe to drink." A majority also agreed that each of the required statements would make them "worried about their drinking water" and that getting the information would make them "change their drinking water habits." The current versions of the USEPA required health effects language for arsenic at the higher levels (10 to 50 µg/L) and the required educational statement for arsenic at the lower levels (5 to 10 µg/L) evoke basically the same responses from consumers, despite the differences in content.

Efforts to clarify the health effects wording resulted in some significant changes in customer ratings. Individuals judged the revised explanation as being more important, clear, informative, worrisome, and likely to cause them to change their drinking water habits. Efforts to clarify the educational statement language by removing mention of a cost-health tradeoff did not seem to have a significant impact on how important, clear, informative, or worrisome the educational statement was perceived to be.

Individuals' responses to the seven explanations with respect to agreement to the five statements are also dependent on their own situation — their satisfaction and trust with their own utility, the perceptions of the adequacy of environmental regulations, and to a lesser extent their sociodemographic characteristics. In particular, individuals who have greater trust in their own water utility are more likely to agree that the statements are important, clear, and informative, and, conversely, individuals who have less trust are less likely to agree.

ADDITIONAL OR ALTERNATIVE COMMUNICATION APPROACHES

This research effort yields the following observations:

- Utilities should consider providing "actionable information," including suggestions such as what one can do to protect against the effects of arsenic in drinking water and information for individuals to find out more about arsenic in their drinking water.
- Information on the utility's arsenic mitigation plans should be provided along with the health effects language or educational statement.

- Utilities with arsenic levels between 5 and 10 μg/L should be prepared to respond to the public just as if they were in violation of the MCL. Simply reinforcing the statement that they are below the MCL is not likely to resolve concerns that customers have.
- Water utilities should consider using focus groups or formal or informal surveys of utility communication efforts to find out more about how their own customers respond to their particular arsenic problem and potential communication efforts on this problem. This approach is relevant not only for communication on arsenic issues but also as part of the utility's overall public communication and customer satisfaction program.

CHAPTER 1
INTRODUCTION

As required in Title 40 Code of Federal Regulations 141.152 (40 CFR 141.152), all community water systems (CWS) must deliver a consumer confidence report (CCR) to their customers by July 1 of each year. CCRs are the centerpiece of the public right-to-know in the federal Safe Drinking Water Act (SDWA). The information contained in CCRs is intended to raise consumers' awareness of where their water comes from, explain the process by which safe drinking water is delivered, and educate consumers about the importance of preventative measures such as source water protection that ensure a safe drinking water supply.

The basic elements of a CCR must include the following:

- Identify the source of water.
- Include boilerplate with standardized phrases and contact information.
- Report contaminants at highest concentration detected.
- Compare the maximum detected concentration of contaminants with the Maximum Contaminant Level (MCL) (explained below).
- Report MCL violations and relevant information about health effects.

The public's health is protected by limiting the levels of specific contaminants in water by setting an enforceable MCL for each contaminant. The MCL is the maximum permissible level of a contaminant in water delivered to any user of a public water system, which includes all public water systems that supply water to at least 25 people and to the same population year-round (USEPA 2002).

In Section 1414(c)(4)(A) of the SDWA Amendments of 1996, Congress required USEPA to establish regulations that provide "brief statements in plain language regarding the health concerns that resulted in the regulation of each regulated contaminant." In subsection (B)(iii)(IV), Congress required CCRs to include this language "for any regulated contaminant for which there has been a violation" of the MCL.

The USEPA established an MCL for arsenic of 10 μg/L[1] in January 2001 [*Federal Register* 2001, 66 (14), 6976].[2] The previous standard for arsenic was 50 μg/L, which was established in 1942. In addition to lowering the arsenic MCL, the USEPA's new arsenic rule created new requirements for public notification and language requirements for inclusion in CCRs. Unlike the new arsenic MCL, which becomes effective in January 2006, these communication requirements became effective February 22, 2002 (see Chapter 2 for a detailed discussion of these requirements).

The visibility of arsenic in the media in recent years, its existence as a natural contaminant, and its established carcinogenic nature make it likely that arsenic will be one of the more identifiable and high profile water contaminants. These factors are also likely to make arsenic of continuing concern for the public once arsenic is addressed in CCRs.

OBJECTIVES

Given the uncertainties of the health effects of drinking water arsenic exposure (Frost et al. 2002; see Chapter 2 for further discussion), and the uncertainties of the public's perception of arsenic risks for low-level exposure in tap water, this research was conducted to help utilities meet the new CCR requirements addressing arsenic. In addition, this work identifies questions and issues about the presence of arsenic in drinking water and the related health risks likely to be raised by the public.

OUTLINE OF REPORT

Chapter 2 discusses the background of arsenic in drinking water and describes the regulations pertaining to arsenic in drinking water and the communication requirements for various arsenic levels. Chapters 3 through 5 discuss specific research tasks undertaken as part of

---

1. One μg/L is roughly equivalent to 1 part per billion; hence a 10 μg/L concentration is roughly equivalent to 10 atoms of arsenic per billion molecules of water.

2. A "Minor Clarification of National Primary Drinking Water Regulation for Arsenic" was published in the Federal Register on March 25, 2003 (68, 57:14501-14507) "to express the new arsenic maximum contaminant level (MCL) as 0.010 mg/L instead of 0.01 mg/L."

this project in conjunction with AwwaRF Project 2692, "Understanding the Impact of CCRs and Emerging Issues Communications on Customers." Chapter 3 discusses a survey given to utilities to elicit input regarding utility concerns with the 2001 Arsenic rule requirements for CCR language. Survey development, implementation, and results are discussed. Chapter 4 includes a discussion of utility customer comments on selected CCRs during focus groups held in five different cities in the United States. Chapter 5 discusses results from a national telephone survey to assess how water utility communications have affected public awareness and perceptions of drinking water quality and safety. Chapter 6 presents the conclusions of the study and recommendations to utilities on how they might address the new arsenic requirements in their CCRs.

CHAPTER 2
ARSENIC IN DRINKING WATER

BACKGROUND

Arsenic is a common, naturally occurring element. It is present in detectable levels in roughly half of U.S. water supplies. Arsenic typically is found in these waters because the element is present naturally in the rocks and soils through which the water passes. However, in some instances, arsenic may be present in drinking water because of nearby mining activities, waste disposal practices, or other sources of pollution.

Arsenic has long been known to be acutely toxic to humans at high doses, and has been immortalized as the poison of choice in classic novels such as *Madame Bovary* and the play *Arsenic and Old Lace.* The levels of arsenic found in U.S. drinking waters are, fortunately, far below the acutely lethal doses (the lethal dose is about 100,000 times higher than the new MCL).[3]

The recent regulatory action has focused on very low levels of arsenic in water, and the possible health consequences of concern are related to long-term ("chronic") exposures to these sublethal levels of arsenic (e.g., decades of exposure to arsenic levels in the parts per billion range). Instead of immediate death, the public health concerns now focus on low levels of potentially elevated risks of cancer (most notably, cancers of the bladder, lung, and skin). Arsenic may also be associated with other possible adverse health effects, including heart disease.

Considerable uncertainty and scientific debate exist about the level of risk posed by arsenic at the levels found in U.S. drinking water. The concern over possible cancer risks from long-term arsenic exposures is based on scientific evidence compiled predominantly in rural Taiwan, South Asia (India and Bangladesh), and South America (Chile). However, the people studied in these locations had been drinking water with arsenic concentrations that typically are 50 times higher than applicable in the United States under the new MCL. Hence, a key question

3. The acutely lethal dose in humans is 70 to 180 mg (Vallee et al. 1960, as cited in Casarett and Doull 1986). This is equivalent to 70,000 to 180,000 µg/L in drinking water, assuming a mean tap water consumption of approximately 1 L/day.

is how the risks observed at high exposure levels translate to risks that may arise at the much lower levels of arsenic found in U.S. waters.

A second unresolved question is how the risks observed in rural Taiwanese and people in other developing nations translate to Americans. U.S. residents typically receive better health care, enjoy more balanced nutrition, and have several other factors that suggest that arsenic risks in the United States would be lower than risks observed among poorer people in developing nations. For example, the rural Taiwanese study subjects differ from most Americans in several potentially important ways, including diet (having higher food-related arsenic exposures and lower levels of other compounds that appear to counteract arsenic toxicity, e.g., selenium), lower incomes (poorer access to health care services and lower baseline health status), and higher smoking rates. To help answer this question, USEPA cosponsored a study of health risks in a Utah population known to be exposed to elevated levels of arsenic in water. The results did not reveal any elevated cancer risks from arsenic exposures in the U.S. study population, but expert reviewers called for additional research before making any conclusive inferences from the Utah data.

The National Research Council (NRC) assembled an expert panel that provided a comprehensive and scholarly review of the scientific evidence pertaining to these questions of how the risks observed in Taiwan (and elsewhere) at arsenic levels of well over 100 μg/L translate to risks applicable to U.S. arsenic levels below the then-current federal MCL of 50 μg/L. The panel did not resolve the major uncertainties, nor did it recommend a specific MCL. However, the panel did reach a consensus that the risk posed at the then current U.S. standard of 50 μg/L may be unacceptably high, and that the MCL should be made more stringent (NRC 1999). Another NRC panel later reviewed and confirmed this finding (NRC 2001). Following the report from the 2001 panel, USEPA affirmed its intention to adhere to the January 2001 rulemaking that established the new MCL of 10 μg/L.

Ultimately, there is still considerable uncertainty and debate over how much risk is posed by long-term exposures to low levels of arsenic found in some U.S. drinking waters. The new USEPA standard of 10 μg/L is intended to reflect a balancing of prudent precaution in protecting public health with the potentially high cost of removing arsenic from drinking water.

Several technologies are available for removing much of the arsenic from drinking water, but these treatment technologies are often expensive to install and operate. In many water systems, complying with the new MCL may lead to noticeably higher water bills for consumers. The potential high cost of the regulation — coupled with uncertainties about the level of health risk reduction that would be realized — prompted much of the public debate over whether the new MCL makes good policy sense.

MONITORING FOR ARSENIC IN DRINKING WATER

Water systems are required to monitor their arsenic levels according to a monitoring schedule, described as follows. The initial monitoring for arsenic at each system involves one sample. If this sample is less than the MCL, then one sample is required every three years for groundwater sources, and one sample is required each year for surface waters. If the initial sample exceeds the MCL, then sampling is required quarterly and a running annual average is computed.

If a system exceeds the MCL, the violation is reported to the state regulatory agency, the violation is recorded, and the system is put on a compliance schedule. The violation must be stated in the system's annual CCR to the community. If violations pose health hazards, they must be reported quickly and public notices must be given.

CCR REPORTING REQUIREMENTS FOR ARSENIC

Under USEPA rulemaking, two different health effects statements are required for arsenic: a health effects statement and an educational statement. The use of these statements in CCRs depends on the level of detected (not average) arsenic concentrations in the water.

Health Effects Language Requirements

The final arsenic rule on January 2001 established an MCL for arsenic of 10 μg/L, but systems do not have to comply with the revised MCL until January 2006. For five years, beginning with the CCRs, due by July 1, 2002, systems that have arsenic levels above the new MCL of 10 μg/L must include an arsenic health effects statement. This action is required even though, technically, the systems are not in violation of the regulations. Then, beginning January

23, 2006, the health effects language need to be included in the CCR only if systems are in violation of the revised arsenic MCL of 10 μg/L.

According to USEPA estimates, before 2002, 197 CWS were required to include the arsenic health effects statement in their CCR (those with arsenic levels greater than 50 μg/L). It is estimated that as of July 1, 2002, under the new arsenic rule, 2,388 systems needed to include the arsenic health effects statement in their CCR (those with arsenic levels greater than 10 μg/L), even though just 197 were in fact in violation of the current standard (USEPA 2000).

USEPA indicated that it is requiring inclusion of the health effects language for arsenic exceedences before the compliance date (January 23, 2006) because it believes that it is important to provide customers with the most current understanding of the risk presented by this contaminant as soon as possible after establishing a new standard. This provision provides systems the flexibility to put this health effects information into context and to explain to customers that the system is complying with existing standards.

The required arsenic health effects language is as follows:

> Some people who drink water containing arsenic in excess of the MCL over many years could experience skin damage or problems with their circulatory system, and may have an increased risk of getting cancer.

Additional Educational Statement Requirements

Congress has allowed USEPA to require CCRs to include a health effects statement even when systems were not violating the MCL, but limited USEPA to just three regulated contaminants for which they may do this. Arsenic is one of the three that USEPA has chosen. As explained in the preamble to the CCR rule (63 FR 44512 at 44514), because of concerns about the adequacy of the current MCL, USEPA decided that systems detecting any arsenic levels over 25 μg/L (one-half the current MCL) must include an additional informational statement. This requirement was updated in 2001 by setting the arsenic level that triggers the additional information to 5 μg/L (one-half the revised MCL), because USEPA indicated that it continues to believe that one-half the MCL is an appropriate trigger for special information about certain contaminants. In 1998, 643 systems had arsenic levels greater than 25 μg/L and 5,577 systems had arsenic levels greater than 5 μg/L (USEPA 2000).

The informational statement is prescribed as follows:

> While your drinking water meets EPA's standard for arsenic, it does contain low levels of arsenic. EPA's standard balances the current understanding of arsenic's possible health effects against the costs of removing arsenic from drinking water. EPA continues to research the health effects of low levels of arsenic which is a mineral known to cause cancer in humans at high concentrations and is linked to other health effects such as skin damage and circulatory problems.

According to USEPA (2001), "Systems retain the flexibility, as defined in the existing requirement, to adjust this additional informational statement in consultation with the Primacy Agency." The term "primacy agency" refers to the state or tribal government entity that has jurisdiction over, and primary enforcement responsibility for, public water systems, even if that governmental entity does not have interim or final primary enforcement responsibility for this rule. Where the state or tribe does not have primary enforcement responsibility for public water systems, the term "primacy agency'' refers to the appropriate USEPA regional office.

The Western Coalition of Arid States (WESTCAS) has recommended to USEPA that the information statement not be required or instead that the following statement be required:

> Although your drinking water contains low levels of arsenic, it fully complies with EPA's standard for arsenic. Systems that are not in compliance with the standard must provide the following information: "some people who drink water containing arsenic in excess of the MCL over many years could experience skin damage or problems with their circulatory system, and may have an increased risk of getting cancer." EPA will continue to research the health effects of low levels of arsenic in drinking water as a part of its periodic review of all drinking water standards.

Their suggested statement begins by telling the reader that the arsenic level does indeed meet the USEPA standard, and then continues by including the health effects statement, which they believe is what Congress intended instead of creating a new statement.

Conversations with several primacy agencies and utilities reveal that not many utilities have requested use of alternative language in their CCRs. Utilities appear to do what is easiest and least costly, which is to use the required language as stated. The primacy agency in Colorado stated that they will "review any alternative language requests and see if they are in line with the intent of the EPA, and suggest any edits if needed and then approve the language if the EPA's intent is clear" (Jackie Roberts, CCR Rule Manager for Colorado, Water Quality Control Division, Colorado Department of Public Health and Environment, personal communication, May 2002.) The primacy agency in California has stated that they "would allow language to be added but they would not allow changes to the stated language" (Alexis Milea, Chief, Standards and Technology Unit, California State Department of Health Services, personal communication, May 2002).

Summary of Arsenic Language Requirements for CCRs

Table 2.1 summarizes the reporting requirements for both the health effects language and the informational statement for various arsenic levels across time. An USEPA Quick Reference Guide is included in Appendix A, which offers an alternative summary of the CCR reporting requirements and provides an overview of the rule and the monitoring requirements.

Table 2.1
Detected arsenic levels and corresponding violations and reporting requirements over time

Detected arsenic levels (µg/L)	Pre-2002		2002-2006		2006 +	
	Violation	Language	Violation	Language	Violation	Language
< 5	No violation	None required	No violation	None required	No violation	None required
5-10				ES required		ES required
10-25				HE required	Violation	HE required
25-50						
50+	Violation	HE required	Violation			

HE = Health effects language.
ES = Educational statement.

CHAPTER 3
UTILITY SURVEY

Research for this project involved three tasks: (1) an informal utility survey, (2) focus groups with water utility customers, and (3) a national telephone survey. We report on these tasks in this and the next two chapters. Because the work on arsenic communication is part of a larger, ongoing AwwaRF project, a more complete description of the full range of objectives and outcomes of each of these tasks will be included in the project report for AwwaRF Project 2692, "Understanding the Impact of CCRs and Emerging Issues Communications on Customers."

OBJECTIVES

In conjunction with ongoing work on AwwaRF Project 2692, drinking water utilities were contacted as part of an informal survey. The survey elicited input from utilities regarding their concerns with the 2001 Arsenic rule requirements for CCR language and explored approaches used by different utilities in meeting these requirements. CCRs were obtained from the surveyed utilities and the language used in these CCRs was characterized.

The survey was designed to determine the communications objectives of the utilities and to identify the types of information most useful for improving communications with their customers. The survey was initially conceived as a short (e.g., 8 or 10 questions) questionnaire designed primarily to guide development of subsequent materials and research methods in a way that would best address the project goals and objectives. Based on findings from the literature review and initial survey design phases of the project, the survey was expanded to capture more detailed insights from the water utilities, as well as to address AwwaRF's request for information about the utilities' experience with and perceptions of USEPA's recent regulation requiring the inclusion of specific arsenic information in CCRs.[4]

4. The marginal cost of adding additional questions to the survey was relatively small given the overall effort required to contact a sample of utilities with the shorter survey.

SURVEY DEVELOPMENT

The utility survey was developed through an iterative process of design and review, and was based, in part, on prior surveys developed for prior AwwaRF research projects. The initial survey instrument was reviewed by the participating utilities,[5] and the survey was revised accordingly. The Project Advisory Committee and AwwaRF Project Manager then reviewed the revised instrument. Finally, the survey was reviewed and revised based on feedback and discussion within the project team.

SURVEY IMPLEMENTATION

The survey questionnaire was mailed to a sample of 401 drinking water utilities in June 2002. The sample was selected by creating a database of CWS, selecting an interim sample of water systems from this database, comparing this sample to the universe in the database, and selecting a final sample of water systems for the mail survey. The CWS data were sorted into 20 groups based on size category (very small, small, medium, large, very large), source water type (surface water, groundwater), and whether a system has had any MCL violations since 1993. After sorting, utilities were randomly selected from each group. There was one exception — only 14 very large groundwater systems had violations and so all 14 were retained. Each utility was contacted by telephone to verify mailing information and to identify the appropriate contact person within the utility.

A survey packet (containing a cover letter, the survey instrument, a flier to remind respondents of key survey requirements, and a prepaid return envelope) was then mailed to each of the 401 selected CWS, including a request for two copies of the utility's most recent CCR.

Completed surveys were returned by 118 utilities for an overall response rate of 30%.[6] Large and very large CWS responded at a higher rate (36% and 48.6%, respectively) than did the

5. The Arsenic project does not formally have participating utilities. These participating utilities are those assisting in Project 2692. These include Austin, Texas; Philadelphia, Pennsylvania; Bellevue, Washington; Tacoma, Washington; Illinois-American (Peoria, Illinois); Cincinnati, Ohio; Bridgeport, Connecticut — BHC Company; and Boulder, Colorado.

6. Eighteen surveys were returned by the U.S. Postal Service for having bad addresses. Consequently, this response rate is based on a total of 383 utilities (i.e., 401 - 18 = 383).

smaller utilities (ranging from about 22% to 26%). The response rate for utilities for which we had been able to verify contact information was 34% compared to 26% for the utilities for which we had been unable to verify contact information.

UTILITY SURVEY RESULTS

The utility survey included 25 "primary" questions. Several questions had multiple response categories. Most of the responding utilities answered all questions, but some questions had fewer answers than others. The follow-up or open-ended questions had larger nonresponse rates.

The first set of questions asked utilities for their contact information and general information, including size, type of population areas served, and types of source waters. This general information is useful for determining if common problems exist across size categories, source water types, or population areas.

The next set of questions gathered information on how utilities meet their CCR requirements. It included information on how they prepare their CCRs, how many they prepare, who they send them to, associated costs, content, perceived impacts, and related comments. In this section, utilities were also asked about their perceptions of the CCR language requirements for arsenic.

Many utilities (44%) consider the required arsenic language important for informing customers about the risks of arsenic in water; 39% neither agree or disagree with this statement, and 16% disagree. However, a similar percentage of utilities (47%) believe that the required arsenic language will only scare customers; 38% neither agree or disagree with this statement, and 16% disagree. Tables 3.1 and 3.2 display the frequencies and summary statistics, respectively, for these two questions.

Table 3.1
Question 19: "In general, how much do you agree or disagree with the following comments about water quality and customer communications?"[a]

Comment	Disagree completely (1)	Disagree somewhat (2)	Neither agree or disagree (3)	Agree somewhat (4)	Agree completely (5)	Doesn't apply to my utility	n[c]
The newly required arsenic language is important for informing customers about the risks of arsenic in water.	4%	12%	39%	32%	12%	16[b]	115
The wording of the required arsenic language will only scare customers.	7%	9%	38%	33%	14%	13[b]	116

a. All percentages listed are calculated as the number of responses to the stated category divided by all responses, not including the "doesn't apply to my utility" responses.
b. Respondents to the "does not apply option" had low or nondetected arsenic levels.
c. Throughout this report "n" is used to indicate the number of utilities or individuals who responded to the question.

Table 3.2
Summary statistics for Question 19

Comment	Mean[a]	Median[a]	Std. dev.[a]	n (not including responses for "doesn't apply to my utility")
The newly required arsenic language is important for informing customers about the risks of arsenic in water.	3.4	3	1.0	99
The wording of the required arsenic language will only scare customers.	3.4	3	1.0	103

a. Statistics are based on the 1 through 5 values given in parentheses in the column headings from Table 3.1.

The next section of the survey asked the utilities what their current maximum arsenic levels are, how they discussed health impacts of arsenic, and details about how they meet CCR requirements for arsenic. It also asked utilities about comments they received from customers regarding arsenic.

Table 3.3 displays the reported maximum arsenic levels reported by 75 utilities, in μg/L. It seems reasonable to assume that the 43 utilities that did not respond to this question had no detected arsenic levels. According to the responses, 26% of the utilities have arsenic levels at 0 or nondetect levels, 23% have levels from 0 to less than 1, 15% have levels from 1 to less than 2, 20% have levels from 2 to less than 5, 8% have levels from 5 to less than 10, and 12% have levels over 10 μg/L.

Table 3.3
"What was the maximum reported level of arsenic in your utility's drinking water for 2001?"

Arsenic level (μg/L)	n	Percent of responses (out of 75 responses)[a]
"not applicable" (na)	1	1%
"not detected" or "below detection limit"[b]	5	7%
"less than the MCL"	1	1%
0	14	19%
More than 0 to less than 1	17	23%
1 to less than 2	11	15%
2 to less than 5	15	20%
5 to less than 10	6	8%
10 to less than 25	2	3%
25 to less than 50	3	4%
More than 50	0	0%
Total	75	100%

a. Seventy-five out of 118 utilities answered this question.
b. Detect levels for arsenic vary among states and may be as high as 10 parts per billion.

Table 3.4 displays the reported arsenic levels in an alternative format and compares them with arsenic levels from USEPA data. The percentages of utilities exceeding the indicated levels of arsenic are shown for our sample and as estimated by USEPA (USEPA 2000). The distributions are comparable, suggesting that we received a reasonably representative sample of responses.

Table 3.4
Comparison of utility survey reported arsenic levels with USEPA reported arsenic levels for all systems

Percentage of systems >	Utility survey responses (out of 75 responses)	USEPA data for all utilities[a] (out of 74,615 systems)
2 μg/L	28.0%	25.9%
5 μg/L	9.3%	11.6%
10 μg/L	4.0%	4.6%
25 μg/L	4.0%	1.3%
50 μg/L	0.0%	0.4%

a. Source: USEPA, 2000.

According to revised USEPA regulations, if 2001 testing showed an arsenic level between 5 and 10 μg/L, the utility must provide an educational statement about arsenic in its CCR. Above 10 μg/L, the utility must include the health effects language. We asked utilities if their 2001 CCRs contained an educational or health effects statement about arsenic. Of the 103 utilities that answered the question, 88% (91 utilities) were not required to have an educational or health effects statement about arsenic in their 2001 CCRs. Eight utilities have the exact statement required by USEPA, one utility has the exact statement plus additional wording, and three utilities have an alternative statement to that required by USEPA.

For instance, one utility with a reported arsenic level of 48 μg/L included the health effects language (not the education statement) in their 2001 CCR and provided the following additional text:

> Meeting USEPA's arsenic limit. In January 2001, USEPA revised the Maximum Contaminant Level for arsenic from 50 to 10 Parts Per Billion. After adopting the new standard, USEPA decided to review the decision to ensure that the standard was based on sound science and accurate estimates of costs and benefits. In October 2001, USEPA decided to move forward with implementing the 10 Parts Per Billion standard. All of the water supplied to our customers meets the current standard. The Utility must comply with the new standard in 2006.

> This allows the Utility time to bring the new Drinking Water Project for using river water into service. Blending river water with ground water could reduce the expected initial cost of arsenic removal from $150 million to around $30 to $40 billion. Over time, however, the Utility would still need to build new arsenic removal facilities to meet growing need.
>
> Arsenic occurs naturally in the earth's crust. When rocks, minerals, and soil erode, they release arsenic into groundwater. Arsenic occurs naturally in varying amounts in Albuquerque's drinking water. For information on arsenic in your distribution zone, visit the web page at www.cabq.gov/waterquality or call the Water Quality Information Line at 857-8260.

Another utility that reported that they did not detect any arsenic in their drinking water, provided the following addition text in the 2001 CCR:

> Drinking water and arsenic. Arsenic has been a hotly discussed drinking water issue for several years. The new federal rule lowers the allowable limit of arsenic in public drinking water from 50 to 10 parts per billion (ppb). Public water systems must implement treatment systems to reduce the level to less than 10 ppb. Arsenic has not been found in the Ames water supply at any detectable level.

The four utilities that stated in the utility survey that included additional or alternative wording about arsenic in their CCR had the following comments:

> Our CCR provided additional information on meeting the standard, and we did not change the wording of the required language. The additional information also directed customers to our web-site for a map of arsenic distribution. No approval was required from our primacy agency.

> We sent our CCR to the State Health Department of North Dakota for approval before printing.
>
> The Ohio Environmental Protection Agency reviewed the entire document before the mailing.
>
> We didn't get any approval. Just as a general statement because of media attention: 'Arsenic Update: In 2001 the Environmental Protection Agency issued stricter regulations regarding Arsenic in drinking water. The current standard lists the Maximum Contaminant Level (MCL) for Arsenic as 50 parts per billion (ppb). The new standard, which will become effective in 2006, is 10 ppb. Test of [city name] water performed in 2001 showed no arsenic present at or above 10 ppb, which will be the new standard.'

This last utility reported a maximum arsenic level of 10 μg/L for 2001 in response to the survey question, but had no level reported in its CCR. Their reported level apparently contradicts the statement in their CCR.

Question 23 asked if the utility had received comments or questions about arsenic. Of the 114 utilities answering Question 23, 13 indicated that they had received comments or questions. Table 3.5 displays results on comments utilities have received from customers concerning arsenic: 11% of the utilities received arsenic-related comments from customers (13 utilities out of the 114 utilities that answered the question). Nine of these 13 utilities received a few comments or questions from customers wanting to know about current arsenic levels, one utility had received no comments on this, one had received several comments on this, and one had received many comments on this issue.

Table 3.5
Question 23: "How many comments did you receive about the following issues?"

	No comments or questions[a] (1)	A few comments or questions[a] (2)	Several comments or questions[a] (3)	Many comments or questions[a] (4)	N[b]
Want to know about current arsenic levels	8%	75%	8%	8%	12
Want information about arsenic health effects	71%	29%	0%	0%	7

a. Values reflect percentage of responses out of the number of responses listed in the final column.
b This question was asked only of the 13 utilities that stated they received comments or questions concerning arsenic.

Two utilities received a few comments or questions from customers wanting information about arsenic health effects, and five received no comments or questions regarding health effects. One utility also reported that customers asked them if fluoride adds arsenic to the water, and one utility reported that customers wanted to know if it samples for arsenic.

The final section of the survey listed a set of potential issues utilities might have to face in the future and asked utilities which ones were potential concerns to them; it also asked them to discuss details concerning the major concerns they stated they might have to face. The top three future concerns of water utilities are infrastructure replacement, changing regulatory requirements, and water supply shortage from both drought and increasing demands. Only one utility of the 118 stated that the arsenic MCL was one of their top future concerns. This utility reported an arsenic level of 10 μg/L.

Utilities were asked to return two copies of their most recent CCRs with the survey. An examination of arsenic-related text in the 117 CCRs received revealed that 41 utilities listed the tested level of arsenic in their detected contaminants table: 49% had levels less than 2 μg/L, 42% had levels greater than 2 and less than 10 μg/L, and 10% had levels over 10 μg/L. These findings are similar to the results from the survey listed previously where 62% of the utilities stated that they had levels less than 2 μg/L, 28% had levels greater than 2 and less than 10 μg/L, and 7% had levels over 10 μg/L.

Eleven of the CCRs examined included an educational statement on arsenic or a discussion of the health effects of arsenic, in accordance with the statements from 12 utilities that they included an educational or health statement.

UTILITY SURVEY CONCLUSIONS

To date, arsenic levels and arsenic reporting do not appear to be a major concern of the utilities surveyed for this work. Given that, many utilities (44%) do consider the required arsenic language important for informing customers about the risks of arsenic in water. However, a similar percentage of utilities (47%) believe that the required arsenic language will only scare customers.

None of the utilities surveyed has exceeded the current arsenic MCL of 50 μg/L; therefore, none of these utilities has needed to include an arsenic health effects statement in its CCR before. However, with the new arsenic rule, 4% of the utilities must now include the health effects statement because their arsenic levels are over 10 μg/L. The distribution of arsenic levels in the utility survey sample is similar to that estimated for all utilities nationally.

About 4% of the utilities reported arsenic levels over 25 μg/L, and thus these 4% were the only ones that needed to include the educational statement. With the new arsenic rule, 9.3% of the utilities would need to include the educational statement because their arsenic levels are greater than 5 μg/L.

A total of 12% of the utilities surveyed included arsenic text in their 2001 CCRs. Most of these utilities included the exact statement required by USEPA, three used an alternative statement, and one provided additional wording along with the required statement directing customers to a website where further information could be found.

Roughly 1 of every 10 utilities reported receiving customer comments concerning arsenic. Most of these customers wanted to know about current arsenic levels, and a few wanted to know about the health effects of arsenic.

CHAPTER 4
FOCUS GROUPS

OBJECTIVES

In July and August 2002, the research team conducted 11 group interviews, or focus groups, with residential utility customers in five cities.[7] The objective of the interviews was to elicit evaluative commentary from customers as they read and reacted to CCRs. This information was a vital precursor of, and supplement to, the national telephone survey (Chapter 5), because it provided important information for survey development and added qualitative depth to numerical ratings obtained in the survey.

PROCEDURES

The focus group question guide was developed to elicit responses from participants in a manner similar to those they may have when receiving a CCR in the mail at home. The sessions were thus as open-ended as possible to allow spontaneous discussions that would elicit individuals' reactions to the CCRs presented and probing regarding their understanding of the information in those CCRs.

Before the interviews, the question guide was pretested in Madison, Wisconsin, with six participants. This exercise resulted in several small refinements of the guide — for instance, two questions were found to be redundant — but fundamentally it confirmed that the questions were meaningful and would produce useful information.

The interviews were conducted in Boulder, Colorado (three group interviews), Philadelphia, Pennsylvania (two group interviews), Bellevue and Tacoma, Washington (two group interviews in each city), and Madison, Wisconsin (two group interviews). Eight participants were recruited for each interview, with an attendance rate of approximately 80% (an

7. Drs. Rambo and Baumgartner of PA Consulting and Dr. Herrick of Stratus Consulting directed the focus group survey research task for AwwaRF Consumer Confidence Report Project 2692 but were not specifically responsible or involved with the work reported here on arsenic issues. The materials reported in this chapter are in part taken from and build on their work on the Consumer Confidence Report Project. They take no responsibility for the conclusions, opinions, or statements of fact expressed in this chapter.

average of 6.5 attendees per interview). The participants were homeowners who received water service from the local utility. An effort was made to obtain a roughly representative mixture of ages and education levels, and approximately half were male and half were female. Participants were paid an incentive of $60 for completing the interview.

Each group was asked to read and comment on the CCR produced by its own utility, and either one or two CCRs from other cities. Table 4.1 shows the pattern of review among all cities.

Table 4.1
Schedule of CCR review

Interview city	First review	Second review	Third review
Boulder (7/17/02)			
First session	Boulder	Bellevue	na
Second session	Boulder	Philadelphia	na
Third session	Boulder	Madison	na
Philadelphia (7/25/02)			
First session	Philadelphia	Boulder	na
Second session	Philadelphia	Bellevue	na
Tacoma (8/6/02)			
First Session	Philadelphia	Tacoma	na
Second session	Madison	Tacoma	Nashville
Bellevue (8/7/02)			
First Session	Nashville	Bellevue	Madison
Second Session	Nashville	Bellevue	na
Madison (8/14 and 8/15/02)			
First session	Philadelphia	Madison	Nashville
Second session	Boulder	Madison	Tacoma

IMPLEMENTATION

The interviews began with questions about local water quality, probing on the topics of taste and odor, color and clarity, and concerns about the healthfulness of the water. Next, participants were asked about the information they receive regarding water quality, and whether they are satisfied with that information. Participants were asked whether they had received a report about the quality of their water. They were asked to write down questions they would like to have answered about their water.

The interview participants were given time to review the first CCR and several aspects of the CCR's design and content were discussed. The second and third CCRs were subjected to essentially the same procedure, although the discussion naturally took on a comparative dimension during these reviews.

The final portion of each interview discussed the participants' reaction to the new mandatory arsenic language. First the language for water that fails to meet the standard above 10 μg/L was read, and reactions recorded. Next, the language for arsenic concentrations between 5 and 10 μg/L was read, and reactions recorded. Additionally, in some instances alternative language, developed from participants' reactions in the Boulder groups, was read. To test whether arsenic is subject to particularly intense concern, in several groups the arsenic language was read, but with trichlorethylene substituted as the contaminant.

FOCUS GROUP RESULTS

The people interviewed for this project were, for the most part, satisfied with their water quality and with the information they received from their water utility regarding their water. The latter was true even though the majority of participants did not recall receiving a water quality report. Some people were eager to have as much information as possible but the majority said they primarily wanted to be informed when there is a problem.

When participants were asked what information about their water they would like to have, the two most common responses were variations on:

- What's in it?
- Is it safe?

Other questions raised about drinking water were:

- How and how often is it tested?
- Where does it come from?
- How is our drinking water processed?
- What affects the taste of our drinking water?

- Is there a change in water quality as it is transported to the home?
- How does local water quality compare to other locations?
- How effective are home filtration systems?

Reactions to the required arsenic language were mixed. Some participants were not alarmed by it at all. A few participants indicated that they were aware that the standard had recently been lowered from 50 μg/L, and some felt the old standard had protected people adequately. Others, however, did find the language troubling. They were especially concerned about the language's uncertainty over the level of risk.

For a number of participants, the wording for the 5-10 μg/L levels was more troubling than the wording for the 10+ μg/L level because the statement reveals that the decision to retain some level of arsenic is based partly on cost issues. They indicated that they felt that placing a price on human health was disturbing or unacceptable.

Some participants said after reading the arsenic statement that they would stop drinking the water, while others said they might minimize use. Some respondents said they would take some action in response to the statement, such as calling the water utility or signing a petition.

Several words and phrases were unclear to participants and raised questions or concerns, including "some people," "in excess," and "circulatory problems." Participants wanted to know what the words really mean and how they could safely use the water. Risk communication research indicates that unclear or ill-defined words or concepts are problematic in risk communication. These general comments suggest that the basic arsenic statements may not provide all components necessary for an adequate risk communication message.

The required arsenic language again for levels above 10 μg/L is as follows:

> Some people who drink water containing arsenic in excess of the MCL over many years could experience skin damage or problems with their circulatory system, and may have an increased risk of getting cancer.

For comparison, alternative language was read to five of the groups. Based on a discussion in the Boulder focus groups, it was intended that this language was more direct in clarifying that the utility's water was in violation of EPA standards:

> Your drinking water does not meet EPA standards for arsenic. It contains arsenic over the levels of acceptability. Therefore, people who drink the water over the course of years may experience increased risks of cancer, skin damage, and circulatory problems.

In most of the groups, individuals preferred this wording for being more honest, though not less troubling. However, when the participants in Philadelphia were read the alternative arsenic statement, they did not prefer it over USEPA's statement. They said the USEPA statement is more "truthful" and does not attempt to cover up the truth.

In some groups, another health effects statement was read for another contaminant to begin to explore whether people have specific concerns about arsenic due to prior knowledge about its toxic effects. Health effects language on trihalomethanes and chlorine dioxide was read and discussed following discussion of the arsenic language. This language was taken largely from the USEPA mandated health effects language for these contaminants:

> Some people who drink water containing trihalomethanes in excess of the MCL over many years could experience skin damage or problems with their liver, kidneys, or central nervous system, and may have an increased risk of getting cancer.

> Some infants and young children who drink water containing chlorine dioxide in excess of the MCL over many years could experience nervous system effects. Similar effects may occur in fetuses of pregnant women who drink water containing chlorine dioxide in excess of the MCL. Some people may experience anemia.

In general, individuals had not heard of these contaminants before this discussion. There was some concern regarding these, that they simply "sound nasty" and are even difficult to pronounce. Additional concern was raised about the specific health effects such as impacts on the central nervous system and risks to fetuses. Because of the possibility of reaction to perceptions of the contaminant in addition to the content of the health effects wording, we included a comparison of the arsenic health effects language to other contaminants in the national telephone survey to explore these effects quantitatively.

A variety of additional issues or thoughts were raised throughout the focus groups but did not necessarily represent consistent or universal themes:

- Some participants were not worried after reading the arsenic statement and said with warnings on so many things today, it did not stimulate concern.
- One participant liked the statement that USEPA is continuing to research the health effects, while another participant stated that "we need the facts now."
- One participant stated they would like to know how much it would cost to remove all arsenic from drinking water and how much it would take in an increase in the water bill to cover the cost.
- Some individuals indicated that the statements didn't indicate what they should do about arsenic in their drinking water — in other words, the arsenic statements do not present actionable information.

FOCUS GROUP CONCLUSIONS

The 11 focus groups held in five cities during the summer of 2002 included homeowners who received water service from the local utility. The participants discussed local water quality and the information they receive about their water, and they examined CCRs from their own utility and at least one other city. They also reviewed and commented on the required arsenic language.

The people interviewed were mostly satisfied with the information they receive from their water utility, even though the majority did not recall receiving a water quality report. Participants discussed the types of information they would like to receive, and most frequently commented they would like to know what is in the water and if their water is safe.

Some participants were not alarmed by the arsenic statement for the 10+ μg/L levels, while most stated the language would cause concern. Several commented the language would leave them confused and cause them to worry, reduce consumption of their tap water, and lose trust in their utility. Some individuals were especially concerned about the language's uncertainty over the level of risk. Particular phrases that were unclear include "some people," "in excess," and "circulatory problems." Participants also wanted more information on how they could safely use the water.

Several participants were concerned about their water after reading the statement for the 5-10 μg/L levels, but were a little less concerned than they were after reading the wording for the 10+ μg/L level. However, they were clearly disturbed that cost was included in the statement as an issue that affected their safety.

A few alternative wordings (including wordings for other contaminants) were read to five of the focus groups, but there were no common reactions shared by a majority. A comparison of the arsenic language to language for other contaminants was included in the national telephone survey to further explore this subject.

CHAPTER 5

NATIONAL TELEPHONE SURVEY

OBJECTIVES

This chapter reports on some of the results of the national telephone survey to assess how water utility communications have affected public awareness and perceptions of drinking water quality and safety. A section of the survey focused primarily on issues related to health effects language with emphasis on the arsenic language and alternatives to the mandated arsenic language. The survey was administered by telephone to a random sample of 1,414 U.S. adult residents nationwide in December 2002.

SURVEY DEVELOPMENT AND IMPLEMENTATION

The survey was developed using input from the utility survey, the focus groups, and prior survey work conducted for AwwaRF by Dr. Robert Hurd (Hurd 1993, Hurd and Becker 1998).[8] The range of topics included questions about the individuals' sources of water, satisfaction with water quality and the water utility (where applicable since about 20% of respondents were on well water), perceptions of information on water quality (including CCRs), and issues related to point-of-use devices and unconventional water sources.

A formal pretest of the survey was conducted with a random sample of roughly 50 respondents before full-scale interviewing to identify and correct any possible problems with the survey questions, sampling, or interviewing. No major problems were identified and the results were included in the final sample total.

To cover a wider range of topics in greater depth, two different versions of the survey were developed and administered to two independent random samples of roughly 700 respondents each (phase one and phase two). Each sample, or phase, included about 550

8. Dr. Hurd directed the national telephone survey research task for AwwaRF Consumer Confidence Report Project 2692 but was not specifically responsible for the work reported here on arsenic issues. The materials reported in this chapter are in part taken from and build on Dr. Hurd's work on the Consumer Confidence Report Project. Dr. Hurd bears no responsibility for the conclusions, opinions, or statements of fact expressed in this chapter.

customers or consumers who get their tap water from a local water utility, and about 150 people who rely on their own private well.

Results from the two samples were combined for analysis purposes, creating a single random sample of 1,414 adults nationwide (1,113 water utility customers and 301 well users) that was representative of the total population and relevant subgroups. A randomly selected subgroup of this sample answered questions relevant to the arsenic health effects language.

The demographics of the total survey sample compared favorably with known population parameters based on current U.S. census figures and previous national surveys conducted by Dr. Hurd for AwwaRF.

Interviews averaged between 8 and 15 minutes, depending on the type of respondent and the number of questions that applied to each group. The survey was administered by experienced, trained professionals from Woelfel Research, Inc. and its telephone interviewing centers in Alexandria, Virginia, and Harrisburg, Pennsylvania.

The survey results from the total sample of 1,414 adults nationwide are accurate within plus or minus 2.6 percentage points — the margin of sampling error allowed for a sample of this size (n = 1,414) at the 95 percent confidence limit.

EVALUATION OF HEALTH EFFECTS LANGUAGE IN THE NATIONAL TELEPHONE SURVEY

Participants in the national telephone survey were read the following statement:

> As I mentioned before, the EPA requires that most water utilities test the local drinking water for many different types of possible contaminants to make sure they do not exceed the Maximum Contaminant Level, or MCL. This is the highest level that is still considered safe and allowed by the federal government. When certain types of contaminants are detected, water utilities are required to provide customers with information on how it could affect your health.

They were then read one of seven explanations on the health effects of various potential contaminants in drinking water. Several of these correspond exactly with USEPA health effects language. Some are modifications of this required language. Table 5.1 shows the seven explanations.

Table 5.1
Selected health explanations for various contaminants

Explanation	Contaminant: explanation	Explanation text
A	Arsenic: health effects	Some people who drink water containing arsenic in excess of the MCL over many years could experience skin damage or problems with their circulatory system, and may have an increased risk of getting cancer.
B	Arsenic: educational statement	While your drinking water meets EPA's standard for arsenic, it does contain low levels of arsenic. EPA's standard balances the current understanding of possible health effects against the costs of removing arsenic from drinking water. EPA continues to research the health effects of low levels of arsenic which is a mineral known to cause cancer in humans at high concentrations and is linked to other health effects such as skin damage and circulatory problems.
C	Arsenic: health effects alternative	Your drinking water contains higher levels of arsenic than considered safe or acceptable by the federal government. Studies suggest that after many years, people who drink water with high levels of arsenic have an increased risk of getting cancer, experiencing skin damage, or developing problems with their circulatory system.
D	Arsenic: educational statement alternative	While your drinking water meets EPA's standard for arsenic, it does contain low levels of arsenic. EPA continues to research the health effects of low levels of arsenic which is a mineral known to cause cancer in humans at high concentrations and is linked to other health effects such as skin damage and circulatory problems.
E	Total coliform bacteria: health effects	Coliforms are bacteria which are naturally present in the environment and are used as an indicator that other potentially harmful bacteria may be present. Coliforms were found in more samples than allowed and this was a warning of potential problems.
F	Total trihalomethanes (THMs): health effects	Some people who drink water containing trihalomethanes in excess of the Maximum Contaminant Level over many years may experience problems with their liver, kidneys, or central nervous system, and may have an increased risk of getting cancer.
G	Chlorine dioxide: health effects	Some infants and young children who drink water containing chlorine dioxide in excess of the Maximum Contaminant Level could experience nervous system effects. Similar effects may occur in fetuses of pregnant women who drink water containing chlorine dioxide in excess of the MCL. Some people may experience anemia.

After hearing one of these explanations, respondents were asked their degree of agreement or disagreement with five statements regarding the health effects language they had just heard. A four-point agreement/disagreement scale was used with a 1 for strongly agree, 2 for somewhat agree, 3 for somewhat disagree, and 4 for strongly disagree. Table 5.2 shows the five statements that respondents responded to. The terms in the "concept" column are used as shorthand for that statement in the discussion and tables that follow on the analysis of responses.

Table 5.2
Statements for evaluating health Explanations A through G

Statement read to respondents	Concept
The statement provided information that was important and worthwhile.	Important
The statement was clear and understandable.	Clear
The statement gave you enough information to decide whether or not your water is safe to drink.	Informative
The statement would make your worried about drinking the water.	Worried
Getting that statement in your local water quality report would make you change your drinking water habits.	Habits

After recording responses on these five scales, respondents were read a different one of the other seven health effects explanations and then asked to indicate their degree of agreement with the same five statements on the same four point agreement/disagreement scale.

TELEPHONE SURVEY RESULTS

Individuals' responses to the various health effects explanation on the various statements were evaluated to see if there are impacts of different explanation or different contaminants on how individuals react to such language as it may appear in CCRs. The results reported here are from the unweighted survey data.[9]

9. Some other analyses from this survey were weighted to correct for the minor deviations in demographics based on census data and/or previous surveys. However, analysis with weighted data would not change any of the substantive conclusions of this chapter.

Overall Responses

Table 5.3 shows the mean responses to the five statements for evaluating health explanations. We look at this first, without distinguishing between the particular contaminant individuals were responding to, in order to see if there is any general pattern in how individuals respond to health effects statements in CCRs. In general individuals tended to agree with the statements (all of the means are less than 2.0, thus falling between "somewhat agree" and "strongly agree").

Table 5.3
Overall responses across all statements

Concept	Mean (standard deviation) Number of observations
Important	1.46 (0.72) 897
Clear	1.50 (0.73) 897
Informative	1.93 (1.00) 861
Worried	1.71 (0.90) 876
Habits	1.76 (0.95) 883

Where 1 = "strongly agree" and 4 = "strongly disagree."

The most disagreement was generated with the "statement gave you enough information to decide whether or not your water is safe to drink," and the most agreement with the "statement provided information that was important and worthwhile." This suggests that while the individuals found it useful and important to receive information on the health risks of these contaminants, were relatively less likely to perceive that they got enough information. Customers perceive this is the area with the most room for improvement.

Comparing Explanations A and B

Explanation A is the required health effects language for arsenic for levels exceeding the MCL (10 μg/L). Explanation B is the "educational statement" for arsenic when tests show its level ranges between one-half the MCL (5 μg/L) and the MCL (10 μg/L).

Explanation B for the lower levels of arsenic was assumed to be less worrisome than Explanation A for higher levels of arsenic. But in the focus groups, it appeared that the opposite might actually be true, due to consumer perceptions of differences in the language or wording.

Since both explanations are mandated language, it was decided to examine whether both explanations are "clear." However, if one was perceived to be more "worrisome," it might also be perceived as more "important" information and more likely to make people "change" their drinking habits. In that case, it is also possible that consumers might be less likely to say the explanation gave them "enough" information.

These propositions were tested using a series of t-tests designed to detect significant differences between pairs of ratings — in this case the ratings respondents gave to the two different explanations.

Table 5.4 shows the summary statistics on individual's responses to the various statements across all individuals who were read Explanations A or B, including the mean responses (rounded to two significant digits) and the Z-statistic, a test of whether there is a significant difference in responses to the statement between the two explanations. A significant difference would suggest that something in the wording of the explanation generates a different response in individuals on that statement.

Table 5.4
Comparing Explanations A and B
Mean
(standard deviation)
n

Statement	Explanation A: Arsenic health effects	Explanation B: Arsenic educational statement	Z-statistic[a] [two-sided probability > Z]
Important	1.50 (0.79) 231	1.50 (0.73) 107	0.175 [0.861]
Clear	1.47 (0.71) 227	1.47 (0.76) 107	-0.297 [0.767]
Informative	1.91 (1.04) 220	1.87 (0.96) 103	-0.007 [0.995]
Worried	1.65 (0.88) 221	1.74 (0.94) 105	0.826 [0.409]
Habits	1.68 (0.94) 223	1.77 (0.92) 96	0.978 [0.328]

a. Z statistics calculated using a Wilcoxon two-sample test using the NPAR1WAY procedure in SAS.

The statistical tests showed no significant differences in the ratings between Explanations A and B. Explanation B (for the lower level of arsenic) was perceived to be just as important, clear, and sufficient as Explanation A (for the higher level of arsenic). Each explanation received high agreement for importance and clarity, and relatively lower agreement for providing sufficient information. That was true for all of the explanations tested in this research (see Table 5.3).

There were also no significant differences between the two explanations and the ratings consumers gave them for making them "worried" or "change their drinking water habits." In each case, most respondents agreed (at least "somewhat") that the explanations would make them worried and change their drinking water habits. This was also true for all of the explanations tested in this research[10]. This is important to the extent that it indicated individuals

[10] It is always possible, although unlikely, that statistically significant differences might be found with larger sample sizes.

are as likely to change their drinking water habits as much when arsenic is above the MCL as when levels are below the MCL in response to the inclusion of information on arsenic in the CCR.

Comparing Explanations A and C

Explanation C is a hypothetical version of the health effects language that was drafted to be more explicit than Explanation A, the current required version for arsenic at levels higher than 10 μg/L).

The hope was that the new hypothetical version (C) might be perceived by consumers as more important, more clear and understandable, and more informative than the current required language in Explanation A. In addition, explanation C was tested to see if it was more effective in raising concern or worry, and making people want to change their drinking water habits. It was also possible that the explanation could be more clear and sufficient, and reduce any unnecessary or undue consumer worry.

Table 5.5 shows the summary statistics on individuals' responses to the various statements across all individuals who were read Explanations A or C.

Table 5.5
Comparing Explanations A and C
Mean
(standard deviation)
n

Statement	Explanation A: Arsenic health effects	Explanation C: Arsenic health effects alternative	Z-statistic [two-sided probability > Z]
Important	1.50 (0.79) 231	1.34 (0.69) 110	-2.269 ** [0.023]
Clear	1.47 (0.71) 227	1.35 (0.60) 112	-1.387 [0.165]
Informative	1.91 (1.04) 220	1.79 (1.02) 107	-1.124 [0.261]
Worried	1.65 (0.88) 221	1.59 (0.85) 113	-0.505 [0.614]
Habits	1.68 (0.94) 223	1.59 (0.94) 110	-1.119 [0.263]

*, **, and *** indicate significance at the 10%, 5%, and 1% level, respectively.

The statistical tests showed that consumers rated the information in Explanation C as significantly more important than the information in Explanation A, and this difference was statistically significant (Z = -2.27, significant at the 5% level).

There were no significant differences, however, in the perceived "clarity" or sufficiency of that information. Explanation C was judged as slightly more clear and more informative than the current required language in Explanation A.

Although Explanation C appeared to be a little more worrisome than the current required language, this difference was not statistically significant. Explanation C was slightly more likely to make people want to change their drinking water habits than Explanation A, but again, this difference is not statistically significant. Although most of these differences were not statistically significant, it is clear from the overall pattern, individuals judged the revised explanation (Explanation C) as being more important, clear, informative, worrisome, and likely to cause them to change their drinking water habits.

Comparing Explanations B and D

Because of the concern expressed in focus groups about decision makers trading off costs against human health Explanation D was developed. It is a minor revision of Explanation B, which is the language required for arsenic levels between 5 and 10 μg/L. The only difference between the two versions is that Explanation D eliminates any reference to cost as a factor in determining safety standards. The focus groups suggested that the reference to cost in B might make the explanation more complex and more worrisome relative to the language in other statements that does not raise the issue of cost.

Table 5.6 shows the summary statistics on individual's responses to the various statements across all individuals who were read Explanations B or D.

Table 5.6
Comparing explanations B and D
Mean
(standard deviation)
n

Statement	Explanation B: Arsenic educational statement	Explanation D: Arsenic educational statement alternative	Z-statistic [two-sided probability > Z]
Important	1.50 (0.73) 107	1.43 (0.65) 102	-0.466 [0.641]
Clear	1.47 (0.76) 107	1.59 (0.77) 110	-1.528 [0.127]
Informative	1.87 (0.96) 103	1.91 (0.88) 96	0.523 [0.601]
Worried	1.74 (0.94) 105	1.83 (0.95) 101	0.817 [0.414]
Habits	1.77 (0.92) 96	2.02 (1.01) 111	-1.814 * [0.070]

*, **, and *** indicate significance at the 10%, 5%, and 1% level, respectively.

Although the statistical tests showed no significant difference in the amount of importance, clarity, "informativeness," and "worry" generated by the two explanations, the current language in Explanation B was significantly more likely than Explanation D to make people say they would "change their drinking water habits."

The results also showed that consumers perceived Explanation B was somewhat more informative than Explanation D (which dropped the reference to cost) — by agreeing more strongly with Explanation B than Explanation D for providing "enough information to decide whether or not your water is safe to drink." This difference was not statistically significant.

Although D was also rated as somewhat more important and somewhat less clear than Explanation B, these differences were not statistically significant.

Taken as a whole, these results suggest that the current language in Explanation B provides more sufficient information and is more likely to lead to behavior change. And cutting the reference to cost (as was done in Explanation D) would most likely soften the impact of the educational statement for low level arsenic, but also make it less informative in the eyes of consumers deciding whether or not the water is safe to drink.

Comparing Explanations A with E, F, and G (Other Contaminants)

The health effects language for arsenic was also compared with health effects language for other contaminants. As revealed in the focus groups, individuals may have different reactions to different contaminants because of prior knowledge (or lack thereof) about the contaminants or because of specific effects indicated by different statements (e.g., impacts on the nervous system from trihalomethanes). To explore this we included health effects statements for total coliform bacteria, total trihalomethanes (THMs), and chlorine dioxide; a microbe, a disinfection byproduct and a disinfectant respectively. With arsenic classified as an inorganic contaminant, this thus covers several of the different types of drinking water contaminants.

Table 5.7 shows the summary statistics on individuals' responses to the various statements across all individuals who were read Explanations A, E, F, and G. The Kruskal-Wallis Test Chi-Square statistic is a test of whether there is a significant difference in responses to the statement between the four explanations. A significant difference would suggest that something in the wording of the different explanations generates a different response in individuals on that statement.

Table 5.7
Kruskal-Wallis Test results for Explanations A, E, F, and G
Mean
(standard deviation)
n

Statement	Explanation A: Arsenic health effects	Explanation E: Total coliform bacteria health effects	Explanation F: Total THMs health effects	Explanation G: Chlorine dioxide health effects	Kruskal-Wallis Test Chi-sq [Pr> Chi-sq] df = 3
Important	1.50 (0.79) 231	1.51 (0.73) 109	1.51 (0.73) 121	1.40 (0.63) 117	1.81 [0.61]
Clear	1.47 (0.71) 227	1.67 (0.86) 122	1.54 (0.75) 114	1.42 (0.63) 105	6.42* [0.09]
Informative	1.91 (1.04) 220	2.08 (0.96) 119	2.02 (1.02) 107	1.94 (1.07) 109	5.05 [0.17]
Worried	1.65 (0.88) 221	1.82 (0.96) 116	1.66 (0.82) 120	1.73 (0.90) 100	3.04 [0.39]
Habits	1.68 (0.94) 223	1.78 (0.98) 110	1.72 (0.88) 123	1.84 (0.96) 110	2.95 [0.40]

*, **, and *** indicate significance at the 10%, 5%, and 1% level, respectively.

The only significant difference between these statements concerns the clarity of the statements. Individuals tended to rate the statement on total coliform bacteria as significantly less clear than the other statements. We suspect that this vagueness may be related to the lack of mention of an MCL for coliform bacteria. It is therefore unclear in the coliform statement whether or not there has been an actual violation of any safety regulation. The statement that "coliforms were found in more samples than allowed and this was a warning of potential problems" doesn't specify what is meant by "allowed" or what the "potential problems" may be. The suggestion that coliforms are an indicator of other bacteria also indicates that there may not be any specific health effects, whereas the negative effects of the other contaminants are more specific.

Although not statistically significant, arsenic was more likely to induce an individual to change their drinking water habits than the other three contaminants. This may reflect a generally higher awareness of the toxic effects of arsenic.

The lack of significant differences between the other contaminants on the other statements suggests that they present the same level and type of information to the consumer irrespective of any prior concepts consumers had about the specific contaminant. This is not to say that this is sufficient or insufficient information, merely that they present the same qualitative information.

Regression Analysis of Responses to Health Effects Language

Simple regression analysis was used as another approach to analyze how individuals responded to the health effects language on the five statements. Using the response to the statement as the dependent variable, the analysis considered how this response is affected by (1) which health effects statement was read, (2) the individual's experience with drinking water and their perceptions of their own water utility, and (3) the sociodemographic characteristics of the individual.

Table 5.8 shows the independent variables used in the regression analysis: the wording of the question in the national telephone survey is shown in the second column, and the measurement scale in the third column. Summary statistics are shown in the last column.

Because of the large number of explanatory variables, the regression results are presented in three separate tables (Tables 5.9, 5.10, and 5.11). Please note that all of the results in these three tables are from a combined regression for each of the five statements.

Table 5.9 presents regression results for the impact of the explanations on the responses to the five statements. The explanations were coded as dummy variables (1 if that statement were read, 0 if another statement was read). Explanation A is the excluded dummy variable, and thus the responses are all interpreted with respect to the response to Explanation A. Parameter estimates on the other explanations can thus be interpreted as measuring how individuals' reactions to the other statements differed from their reaction to the USEPA required arsenic statement.

Table 5.8
Independent variables in the regression analysis

Variable	Question	Scale	Mean (std. dev.) n
Experience with and perceptions of drinking water			
SATISFIED	Overall, how satisfied are you as a customer with your local water utility?	Totally satisfied = 1; not at all satisfied = 5	2.13 (1.19) 1102
UTILITY JOB	Overall, how would you rate the job your local water utility is doing today?	Excellent = 1; Poor = 4	2.01 (0.88) 1082
FILTER	Other than a water softener, do you use a water purification system or filtering device in your home?	Yes = 1; No = 2	1.65 (0.48) 1114
OWN SAFE	Have you ever seen, read or heard anything in the national or local media that made you think that the water in your own community might be unsafe or unhealthy to drink?	Yes = 1; No = 2	1.76 (0.42) 1088
TRUST	Overall, how much trust do you have in your local water utility?	Totally = 1; Not at all = 5	2.13 (1.03) 1112
REGULATIONS	The federal government currently sets standards or guidelines for the quality and safety of drinking water nationwide. Do you think the federal standards are:	Too strict = 1; Not strict enough = 3	2.46 (0.52) 1118
GET INFO	I am getting the information I need to know whether or not my tap water is safe to drink.	Strongly agree = 1; Strongly disagree = 4	2.12 (1.00) 1118
GOT CCR	Can you remember if your household has ever received [a] water quality reports?	Yes = 1; No = 2	1.56 (0.44) 1118
Sociodemographics			
CHILDREN	Do you have any children under the age of 18 living with you in your household?	Yes = 1; No = 2	1.65 (0.48) 1118
RACE	Which of the following categories best describes your racial background?	White = 1; Other = 0	0.78 (0.42) 1118
GENDER		Male = 1; Female = 0	0.39 (0.49) 1118
INCOME	In 2001, [what] was your total household income?	('000,000s of 2001 US dollars)	0.04 (0.02) 1118

Table 5.9
OLS regression on responses to evaluation statements: Explanations
(standard deviation in parentheses)

	Important	Clear	Informative	Worried	Habits
N	830	831	793	809	814
Adj-R-Sq	0.026	0.059	0.102	0.058	0.065
Intercept	1.018***	0.748***	1.177***	2.407***	2.608***
	(0.265)	(0.276)	(0.368)	(0.341)	(0.356)
Explanation B	0.031	0.013	0.093	0.086	0.147
	(0.084)	(0.088)	(0.118)	(0.110)	(0.117)
Explanation C	-0.185**	-0.117	-0.001	-0.091	-0.044
	(0.083)	(0.086)	(0.117)	(0.107)	(0.112)
Explanation D	-0.099	0.093	0.073	0.178	0.381***
	(0.086)	(0.087)	(0.121)	(0.111)	(0.112)
Explanation E	0.032	0.195**	0.202*	0.174*	0.156
	(0.082)	(0.083)	(0.111)	(0.105)	(0.111)
Explanation F	0.004	0.040	0.128	0.020	0.032
	(0.081)	(0.086)	(0.117)	(0.105)	(0.108)
Explanation G	-0.059	-0.045	0.057	0.068	0.182
	(0.080)	(0.087)	(0.114)	(0.110)	(0.111)

*, **, and *** indicate significance at the 10%, 5%, and 1% level, respectively.

Consistent with results reported earlier comparing Explanations A and B, it is interesting to note that Explanation B was not significantly different on any of the statements from Explanation A. Overall, this suggests that even though Explanation A is for a violation of the MCL and Explanation B is intended as an information statement about a risk below the MCL, both statements have the same effect on individuals.

The significant coefficient on Explanation C for "Importance" indicates that the alternative Explanation C was perceived as less important than the original mandated language. This is the same finding discussed earlier on the comparison of Explanation A and Explanation C.

It is also interesting that Explanation E or total coliform bacteria is significantly different from Explanation A on three of the five statements. Even though individuals perceived that Explanation E was less clear and less informative than the arsenic statement, it also worried them less than the arsenic statement.

Table 5.10 shows the portion of the regression results for experience with and perceptions of drinking water. While the statements were read as not necessarily pertaining to the individual's utility, how an individual currently feels about their own utility or their perceptions of water quality most likely affect how they respond to information about water contaminants.

Table 5.10
OLS regression on responses to evaluation statements: Experience with and perceptions of drinking water
(standard deviation in parentheses)

Dependent variable	Important	Clear	Informative	Worried	Habits
SATISFIED (Totally satisfied = 1; Not at all satisfied = 5)	0.006 (0.031)	-0.068 (0.032)**	-0.138 (0.043)***	0.019 (0.042)	-0.028 (0.042)
UTILITY JOB (Excellent = 1; Poor = 4)	0.013 (0.043)	0.097 (0.044)**	0.093 (0.059)	-0.067 (0.056)	0.023 (0.058)
FILTER (Yes = 1; No = 2)	0.106 (0.052)**	0.043 (0.054)	0.107 (0.073)	0.112 (0.067)*	-0.001 (0.071)
HOW SAFE (Yes = 1; No = 2)	0.096 (0.061)	-0.015 (0.065)	-0.122 (0.086)	-0.023 (0.081)	-0.078 (0.083)
TRUST (Totally = 1; Not very much = 4)	0.113 (0.037)***	0.128 (0.037)***	0.154 (0.050)***	-0.063 (0.047)	-0.104 (0.049)**
REGULATION (Too strict =1; Not strict enough = 3)	-0.105 (0.052)**	0.064 (0.053)	0.115 (0.071)	-0.272 (0.067)	-0.207 (0.067)***
GOT CCR (Yes = 1; No = 2)	0.037 (0.056)	0.094 (0.060)	-0.061 (0.079)	-0.152 (0.074)**	-0.114 (0.076)
GET INFO (Strongly agree = 1; Strongly disagree = 4)	0.005 (0.029)	0.013 (0.030)	0.141 (0.040)***	0.003 (0.038)	-0.040 (0.039)

*, **, and *** indicate significance at the 10%, 5%, and 1% level, respectively.

The most striking finding is that individuals who currently trust their own utility more felt that the statements were significantly more important, clear, and informative than those people who trusted their utility less. In addition, individuals who currently trust their own utility would be less likely to change their habits after reading these statements. This reinforces the importance to a water utility of generating and maintaining trust with their customers.

The significant and negative estimates on SATISFIED for the Clear and Informative statements seem counterintuitive, at first. These suggest that individuals who are less satisfied perceive the information in the statements as being clearer and more informative than individuals who are currently less satisfied with their own utility. However, it should be noted that the messages are all negative to one degree or another.

The positive and significant estimate on GET INFO for the Informative statement suggest that individuals who currently do not feel they get enough information on water quality also felt that they explanations did not provide enough information.

Table 5.11 presents the third portion of the regression results – individuals' responses based their sociodemographic characteristics. The most significant finding here is that males indicated that they thought the explanations were more informative than females did. Such explanations were also less likely to make males worried and would be less likely to make them "change your drinking water habits."

Table 5.11
OLS regression on responses to evaluation statements: Sociodemographics
(standard deviation in parentheses)

Dependent variable	Important	Clear	Informative	Worried	Habits
CHILDREN	-0.025	0.056	-0.157	0.021	0.045
(Yes = 1; No = 2)	(0.052)	(0.054)	(0.074)**	(0.067)	(0.069)
Race	0.049	-0.031	0.099	0.093	0.073
(White = 1; Other = 0)	(0.060)	(0.063)	(0.086)	(0.078)	(0.082)
Gender	0.057	-0.035	-0.143	0.178	0.203
(Male = 1; Female = 0)	(0.050)	(0.051)	(0.070)**	(0.064)***	(0.067)***
Income	-0.282	3.077	8.300	0.134	0.175
('000s 2002 Dollars)	(1.287)	(1.345)**	(1.774)	(1.662)	(1.720)

*, **, and *** indicate significance at the 10%, 5%, and 1% level, respectively.

Overall, the regression results indicate that the wording of the explanations does have an impact on how individuals respond to health effects statements. This was seen especially with the wording used for total coliform bacteria. Individuals' experience with their own water utility and perceptions of water quality also affect how individuals respond to this risk information; in particular, trust plays a large role in how individuals respond. Sociodemographic characteristics also play a role in how individuals respond, especially gender differences.

CONCLUSIONS FROM TELEPHONE SURVEY RESULTS

In each case, a majority agreed that the information is "important and worthwhile" and "clear and understandable." They are somewhat less likely to agree that the statements provide "enough information to decide whether or not your water is safe to drink."

This is probably understandable since respondents were simply read these statements without any of the other supporting information that might be available in a typical water quality report.

A majority also agreed that each of the required statements would make them "worried about their drinking water" and that getting the information would make them "change their drinking water habits."

The current versions of the required health effects language for arsenic at the higher levels (10-50 μg/L) and the required educational statement for arsenic at the lower levels (5-10 μg/L) evoke basically the same responses from consumers. No significant differences were detected — at least in the current language.

Efforts to clarify the health effects wording (i.e., replacing Explanation A with Explanation C) did result in significant changes in customer ratings. Individuals judged the revised explanation as being more important, clear, informative, worrisome, and likely to cause them to change their drinking water habits. Efforts to clarify the educational statement language (i.e., replacing Explanation B with Explanation D) did result in changes in customer ratings, but not necessarily the improvements hoped for. In particular, removing mention of a cost-health tradeoff did not seem to have a significant impact on how important, clear, informative, or worrisome the educational statement was perceived to be. This is not to say that additional information or changes to the existing language is not advised. Overall, the language could be supplemented to improve the content of the risk message.

Only minor differences were found in responses to contaminants other than arsenic. The one significant difference was with respect to responses to the wording for coliform bacteria — wording for this contaminant appeared to be less informative and useful that language for other contaminants.

Individuals' responses to the seven explanations with respect to agreement to the five statements are also dependent on their own situation — their satisfaction and trust with their own utility, the perceptions of the adequacy of environmental regulations, and to a lesser extent their sociodemographic characteristics.

The current and alternate statements are more likely to meet the needs of people who already trust their utility, rather than those who are less sure and presumably have a greater need or desire for information. Utilities are encouraged to take the opportunity provided by the CCR's flexible requirements to cater their communication statements to address the almost 30% of their customers who do not have trust and take advantage of this opportunity to help customers feel better about their utility.

CHAPTER 6
CONCLUSIONS AND RECOMMENDATIONS

According to USEPA estimates (USEPA 2000), over 5,500 systems have to deal with the requirements for arsenic health effects language or educational statements in their CCRs, unless they have mitigation in place to reduce arsenic levels below 5 μg/L. There has been little research on the effects of specific types of language in the CCRs, and very little or no research on the arsenic language specifically. The purpose of this project was to provide guidance to community water systems on how to address CCR and risk communication requirements related to arsenic.

Research on the effect of utility communications on customer satisfaction suggests that in general customers want more information from utilities, especially on safety. The literature also suggests that customers who have read more information have a higher level of satisfaction with and trust in their utility (Hurd, 1993; Hurd, 2003). Trust is important as part of the overall culture of the water utility and, in part, an outcome of its public information program.

As shown in the regression analysis, lack of trust in a utility has a significant impact on how individuals respond to arsenic health effects language: those with less trust consider the information explanations less important, less clear, and less informative than those people who trust their utility more. In addition, individuals who trust their utility less are more likely to change their habits after reading these statements — perhaps an indication that they do not trust the utility to take care of the problem and thus have to act themselves to ensure that their drinking water is safe.

Trust is a quality of the long-term relationship between the utility and the customer and cannot be gained entirely by any single communication effort — although it is possible to lose trust with an errant communication effort. Communications about arsenic thus need to be undertaken in context of the water utility's overall risk communication program (as found for all types of communications in general in Hurd and Becker, 1998).

Communications about arsenic are likely to be an important component of utilities' communication efforts for systems where arsenic levels are not only above the MCL but also above one-half the MCL, triggering the need for the education statement in the CCR. Although not the subject of this research, even systems that undertake significant capital investments to mitigate arsenic levels most likely still have to deal with risk communication issues to explain increased costs to their customers. For instance, communications justifying increased costs still take on the character of risk communication regarding decision making about arsenic.

The required and alternative arsenic language was tested on participants of several focus groups and explored in the national telephone survey. In general, respondents agreed that the explanations were important, worthwhile, clear, and understandable, although they were less likely to agree that the statements provided enough information to help them decide whether or not their water was safe to drink. Respondents also indicated that the explanations would make them worried about their drinking water and that the explanations would make them change their drinking water habits.

Overall results indicated that there was not much difference in the impact between the health effects statements (for arsenic levels over 10 μg/L) and educational statements (for arsenic levels between 5 and 10 μg/L). This is an important finding in itself in that people didn't react differently when told the water was above the MCL (i.e., considered by USEPA to be unsafe) and below the MCL (i.e., considered to be safe).

An alternative to the required health effects statement for arsenic (μg/L > 10) generated some significant changes in participants' responses and overall seemed to generate a more favorable response from individuals. Although only one alternative to the health effects language was explored, utilities might do well to supplement the required health effects language with additional information. Given possible difficulties in obtaining approval for revising the language, additional information may be preferred to trying to obtain approval from the primacy agency to implement alternative language.

Comparing responses to Explanation B, the educational statement, and Explanation D, an alternative to the educational statement, indicated that people may also respond differently to an alternative statement here. In particular in the focus groups, the implicit trading of costs against human health is an important issue to some people. In the terminology of risk communication, this issue generates "outrage" among some members of the public. Such outrage can have a disproportionately large impact on the utility's dealings with the public. Again, even though USEPA allows for alternative language and the process appears to be fairly simple, utilities rarely have altered the required language in the past. It is uncertain whether primacy agencies would allow alternative language, in which case additional language may be in order.

A few simple suggestions follow from this research with respect to what additional or alternative information or communication approaches a utility should consider.

Foremost, utilities should consider providing "actionable information." The basic health effects language and educational statements are not complete risk communication messages. They do not provide readers with information on how to react to the information and thus may leave customers confused and frustrated. Actionable information could include suggestions such as what one can do to do to protect against the effects of arsenic in drinking water, a question raised by several focus group subjects. For instance, some subjects wanted to know whether boiling water would solve the arsenic problem — a perception that utilities would want to correct because boiling water may lead to higher arsenic concentrations.

Actionable information would also include information on how individuals could learn more about arsenic in their drinking water. This could include a phone number for a knowledgeable contact person at the water utility, a link to a web page (such as the web page suggested on cryptosporidium suggested in the AwwaRF project "Protocol for Cryptosporidium Risk Communication"), or contact information at the state health agency or USEPA.

Subjects in focus groups also indicated that they wanted to know what the utility was doing about the problem. It could be appropriate to include information on arsenic mitigation plans along with the health effects language or educational statement (i.e., in the same part of the CCR).

Our findings also suggest that whether arsenic levels above 10 μg/L or between 5 and 10 μg/L, individuals are likely to react largely the same with respect to wanting more information, being worried about the situation, and wanting to change their drinking water behavior. Utilities with arsenic levels between 5 and 10 μg/L should thus be prepared to provide information to the public just as if they were in violation of the MCL. Simply reinforcing the statement that they are below the MCL is not likely to resolve customer concerns.

Finally, water utilities should consider using focus groups or formal or informal surveys on utility communication efforts to find out more about how their own customers respond to their particular arsenic problem and potential communication efforts on this problem. As found in the regression analysis, different individuals respond differently to communication messages, and thus a water utility should better understand their customers and how best to provide them with information. This approach is relevant not only for communication on arsenic issues but also as part of the utility's overall public communication and customer satisfaction program.

APPENDIX A

EPA QUICK REFERENCE GUIDE FOR ARSENIC AND CLARIFICATIONS TO COMPLIANCE AND NEW SOURCE MONITORING RULE

United States Environmental Protection Agency | Office of Water (4606) | EPA 816-F-01-004 January 2001

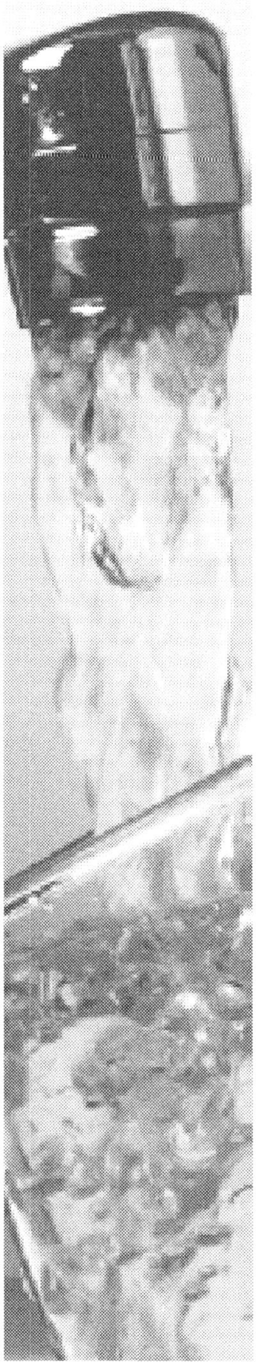

Arsenic and Clarifications to Compliance and New Source Monitoring Rule: A Quick Reference Guide

Overview of the Rule

Title	Arsenic and Clarifications to Compliance and New Source Monitoring Rule 66 FR 6976 (January 22, 2001)
Purpose	To improve public health by reducing exposure to arsenic in drinking water.
General Description	Changes the arsenic MCL from 50 μg/L to 10 μg/L; Sets arsenic MCLG at 0; Requires monitoring for new systems and new drinking water sources; Clarifies the procedures for determining compliance with the MCLs for IOCs, SOCs, and VOCs.
Utilities Covered	All community water systems (CWSs) and nontransient, noncommunity water systems (NTNCWSs) must comply with the arsenic requirements. EPA estimates that 3,024 CWSs and 1,080 NTNCWSs will have to install treatment to comply with the revised MCL.

Public Health Benefits

Implementation of the Arsenic Rule will result in . . .	• Avoidance of 16 to 26 non-fatal bladder and lung cancers per year. • Avoidance of 21 to 30 fatal bladder and lung cancers per year. • Reduction in the frequency of non-carcinogenic diseases.

Critical Deadlines & Requirements

Consumer Confidence Report Requirements *

Report Due	*Report Requirements*
July 1, 2001	For the report covering calendar year 2000, systems that detect arsenic between 25 μg/L and 50 μg/L must include an educational statement in the consumer confidence reports (CCRs).
July 1, 2002 and beyond	For reports covering calendar years 2001 and beyond, systems that detect arsenic between 5 μg/L and 10 μg/L must include an educational statement in the CCRs.
July 1, 2002 - July 1, 2006	For reports covering calendar years 2001 to 2005, systems that detect arsenic between 10 μg/L and 50 μg/L must include a health effects statement in their CCRs.
July 1, 2007 and beyond	For reports covering calendar year 2006 and beyond, systems that are in violation of the arsenic MLC (10 μg/L) must include a health effects statement in their CCRs.

For Drinking Water Systems

Jan. 22, 2004	All *NEW* systems/sources must collect initial monitoring samples for all IOCs, SOCs, and VOCs within a period and frequency determined by the State.
Jan. 1, 2005	When allowed by the State, systems may grandfather data collected after this date.
Jan. 23, 2006	The new arsenic MCL of 10 μg/L becomes effective. All systems must begin monitoring or when allowed by the State, submit data that meets grandfathering requirements.
Dec. 31, 2006	Surface water systems must complete initial monitoring or have a State approved waiver.
Dec. 31, 2007	Ground water systems must complete initial monitoring or have a State approved waiver.

For States

Spring 2001	EPA meets and works with States to explain new rules and requirements and to initiate adoption and implementation activities.
Jan. 22, 2003	State primacy revision applications due.
Jan. 22, 2005	State primacy revision applications due from States that received 2-year extensions.

* For required educational and health effects statements, please see 40 CFR 141.154.

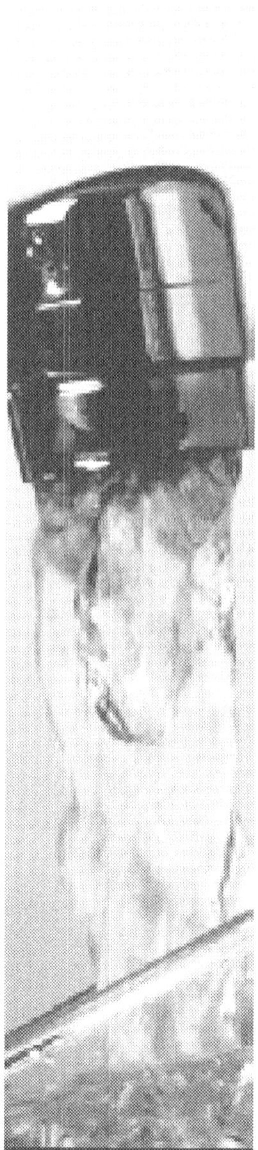

Compliance Determination (IOCs, VOCs, and SOCs)

1. Calculate compliance based on a running annual average at each sampling point.
2. Systems will not be in violation until 1 year of quarterly samples have been collected (unless fewer samples would cause the running annual average to be exceeded.)
3. If a system does not collect all required samples, compliance will be based on the running annual average of the samples collected.

Monitoring Requirements for Total Arsenic [1]

Initial Monitoring

One sample after the effective date of the MCL (January 23, 2006). Surface water systems must take annual samples. Ground water systems must take one sample between 2005 and 2007.

Reduced Monitoring

If the initial monitoring result for arsenic is less than the MCL . . .	Ground water systems must collect one sample every 3 years. Surface water systems must collect annual samples.

Increased Monitoring

A system with a sampling point result above the MCL must collect quarterly samples at that sampling point, until the system is reliably and consistently below the MCL.

[1] All samples must be collected at each entry point to the distribution system, unless otherwise specified by the State.

Applicability of the Standardized Monitoring Framework to Arsenic

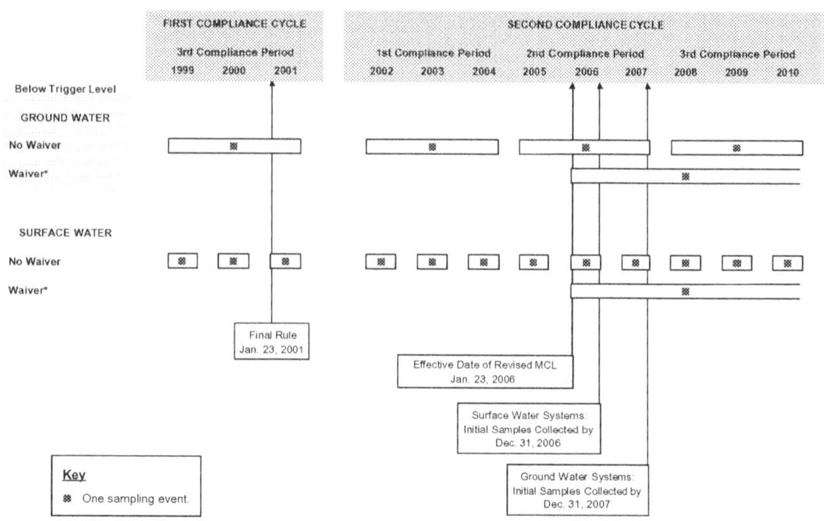

*Waivers are not permitted under the current arsenic requirements. States may issue 9 year monitoring waivers under the revised final arsenic rule. To be eligible for a waiver, surface water systems must have monitored annually for at least 3 years. Ground water systems must conduct a minimum of 3 rounds of monitoring with detection limits below 10 µg/L.

For additional information on the Arsenic Rule

Call the Safe Drinking Water Hotline at 1-800-426-4791; visit the EPA Web site at www.epa.gov/safewater; or contact your State drinking water representative. EPA will provide arsenic training over the next year.

REFERENCES

Casarett, L.J., and J. Doull. 1986. *Toxicology: The Basic Science of Poisons*. 3d ed. New York: Macmillan.

Frost, F., K. Tollestrup, G. Craun, R. Raucher, J. Stomp and J. Chwirka. 2002. Evaluation of Costs and Benefits of a Lower Arsenic MCL. *Journal AWWA*, 94(3):71-80.

Hurd, R.E. 1993. Consumer Attitude Survey on Water Quality Issues. AWWA and the AWWA Research Foundation, Denver, CO.

Hurd, R.E. 2003. Consumer Attitude Survey Update. AWWA Research Foundation. Denver, CO.

Hurd, R.E., and J. Becker. 1998. AWWA Focus Groups to Develop and Test Effective Water Quality Reports: Responding to CCR Requirements in the SDWA. Final Report on WITAF Project No. 408. AWWA, Washington, D.C.

NRC (National Research Council). 1999. Arsenic in Drinking Water. March 1999. National Academy Press, Washington. www.nap.edu.

NRC (National Research Council). 2001. Arsenic in Drinking Water: 2001 Update. National Academy Press, Washington, DC. www.nap.edu.

USEPA (U.S. Environmental Protection Agency). 2000. Arsenic Occurrence in Public Drinking Water Supplies. EPA-815-R-00-023. Prepared by ISSI Consulting Group, The Cadmus Group, and ICF Consulting for the Office of Ground Water and Drinking Water. December.

USEPA (U.S. Environmental Protection Agency. 2001. National Primary Drinking Water Regulations; Arsenic and Clarifications to Compliance and New Source Contaminants Monitoring; Final Rule. Federal Register. 66:14:7033. January 22.

USEPA (U.S. Environmental Protection Agency). 2002. Current Drinking Water Standards. Office of Groundwater and Drinking Water Website. http://www.epa.gov/OGWDW/standard/setting.html. Accessed November 21, 2002.

Vallee, B.L., D.D. Ulmer, and W.E.C. Wacker. 1960. Arsenic Toxicology and Biochemistry. *Arch. Ind. Health*, 21:132-151.

ABBREVIATIONS

CCR	consumer confidence report
CWS	community water systems
MCL	maximum contaminant level
μg/L	micrograms per liter
NRC	National Research Council
ppb	parts per billion
SDWA	Safe Drinking Water Act
THMs	total trihalomethanes
USEPA	U.S. Environmental Protection Agency
WESTCAS	Western Coalition of Arid States

REFERENCES

AWWA. 1998. *Preparing Consumer Confidence Reports*. Denver, Colo.: AWWA.

Bogert, J. 1985. In Defense of the Fog Index. *The Bulletin*, June.

Clark, N., H. Conover, C. Delude, J. Duffy, and K. Mirvis. 2000. Getting out the CCR: A Regional Survey of Water Suppliers. *Jour. NEWWA*, March:32-36.

Colbourne, J., A. Owen, L. Godfrey, J. Brisco, and D. Holt. 2001. *Tools and Methods to Effectively Measure Customer Perceptions*. Denver, Colo.: AwwaRF and AWWA.

CSADW (Campaign for Safe and Affordable Drinking Water). 2000. Measuring Up: Grading the First Round of Drinking Water Right to Know Reports [Online]. Available: <http://www.consumerfed.org/water.pdf>. [cited: June 2002]

Daffron, S. 2003. Improve the Readability of Your Newsletter [Online]. Available: <http://www.newsletterhelp.com/LPMArticle.asp?ID = 2>. [cited January 2003]

Downs, R.M., and D. Stea, eds. 1973. Cognitive Maps and Spatial Behaviour: Process and Product. In *Image and Environment*. London: Edward Arnold.

ETC Institute. [N.d.] *Customer Attitudes, Behavior and the Impact of Communications Efforts*. Denver, Colo.: AwwaRF. Forthcoming.

George Washington University. [N.d.] *Strategies for Identifying Emerging Drinking Water Contaminants and Communicating to Address Public Concerns*. Denver, Colo.: AwwaRF. Forthcoming.

Hurd, R.E. 1993. *Consumer Attitude Survey on Water Quality Issues*. Denver, Colo.: AWWA and AwwaRF.

Hurd, R.E. 1997. Consumer Attitude Survey Update: Preliminary Results From Phase One. Presented at the AWWA Annual Conference, Atlanta, Ga., June 16.

Hurd, R.E. 2003. *Consumer Attitude Survey Update*. Denver, Colo.: AwwaRF.

Hurd, R.E., and J. Becker. 1998. *AWWA Focus Groups to Develop and Test Effective Water Quality Reports: Responding to CCR Requirements in the SDWA*. Final Report on Water Industry Technical Action Fund (WITAF) Project No. 408. Washington, D.C.: AWWA.

Johnson, B.B. 1997. Consumer Confidence Report for Drinking Water Contamination: Initial Studies on Public Response. Presented at the Society for Risk Analysis meeting, Washington, D.C., December 10.

Johnson, B.B. 2000. Using Customers' Views of the "Consumer Confidence Report" of Drinking Water Quality. *Risk: Health, Safety, and Environment*, Fall:309-328.

Johnson, B.B. 2001. Public Reaction to Mandated Language for U.S. Drinking Water Quality Reports. *Risk: Health, Safety, and Environment*, Fall:153-178.

Johnson, B.B. 2002. Comparing Bottled Water and Tap Water: Experiments in Risk Communication. *Risk: Health, Safety, and Environment*, Spring:69-94.

Johnson, B.B. [N.d.] Do Reports on Drinking Water Quality Affect Customers' Concerns? Experiments in Report Content. *Risk Analysis*. Forthcoming.

JohnstonWells. 2001. *Communicating With Customers*. Denver, Colo.: AWWA.

Jones-Lee, M., and G. Loomes. 1995. Scale and Context Effects in the Valuation of Transport Safety. *Journal of Risk and Uncertainty*, 11:183-203.

Jordan, C.G. 1998. Water Quality Reports Boost Consumer Confidence. *Jour. AWWA*, 90(1):40-43.

Kitchin, R., and S. Freundschuh, eds. 2000. *Cognitive Mapping: Past, Present and Future*. New York: Routledge.

Lazo, J.K., J.C. Kinnell, and A. Fisher. 2000. Expert and Layperson Perceptions of Ecosystem Risk. *Risk Analysis*, 20(2):179-193.

Macro International, Inc. 1998. *Focus Groups on Consumer Confidence Reports: Focus Group Report.* Washington, D.C.: USEPA Office of Ground Water and Drinking Water.

Morgan, M.G., B. Fischhoff, C.J. Atman, and A. Bostrom. 2001. Risk Communication: A Mental Models Approach. Cambridge, U.K.: Cambridge University Press.

NRDC. 2003. What's on Tap? Grading Drinking Water in U.S. Cities [Online]. Available: <http://www.nrdc.org/water/drinking/uscities/execsum.asp#>. [cited May 2003]

Oswald, L., and R.E. Hurd. 1998. H2Ohhh! Shaping Perceptions of Water Quality and Trust. Presented at the AWWA Annual Conference, Dallas, Texas, June 23.

Owen, A.J., J.S. Colbourne, C.R.I. Clayton, and C. Fife-Shaw. 1999. Risk Communication of Hazardous Processes Associated With Drinking Water Quality: A Mental Models Approach to Customer Perception, Part 1—A Methodology. *Water Science Technology*, 39(10-11):183-188.

Rambo, E., R. Baumgartner, and C. Koenig. 2003. *Developing Customer Service Targets Through Assessing Customer Perspectives.* Denver, Colo.: AwwaRF.

Rouse, W.B., and N.M. Morris. 1986. On Looking into the Black Box: Prospects and Limits in the Search for Mental Models. *Psychological Bulletin*, 100(3):349-363.

Small, M.J., B. Fischhoff, E.A. Casman, C. Palmgren, and F.W. Morris. 2002. *Protocol for* Cryptosporidium *Risk Communication.* Denver, Colo.: AwwaRF.

U.S. Census. 2000. Current Population Survey, Table 1 [Online]. <http://www.census.gov/prod/2001pubs/p20-537.pdf>. [cited March 27, 2002]

USEPA. 1999. *State Implementation Guidance for the Consumer Confidence Report (CCR) Rule.* EPA 816-R-99-008. Washington, D.C.: USEPA.

USEPA. 2001. 40 CFR Parts 9, 141, and 142. National Primary Drinking Water Regulations; Arsenic and Clarifications to Compliance and New Source Contaminants Monitoring; Final Rule. *Fed. Reg.* (Part VIII), 66(14):6975-7066.

USEPA. 2002. Office of Water. Factoids: Drinking Water and Ground Water Statistics for 2000 [Online]. Available: <http://www.epa.gov/safewater/pws/factoids.html>. [cited September 2002]

USEPA. 2003a. Consumer Confidence Reports [Online]. Available: <www.epa.gov/safewater/ccr1.html>. [cited January 2003]

USEPA. 2003b. CCR iWriter Web page [Online]. Available: <www.ccriwriter.com>. [cited January 2003]

Young, M., S. Dahan, and M. Shaffer. 2001. *Measuring Up II: An Evaluation of Water Quality Information Provided to Drinking Water Consumers in California.* San Francisco: Clean Water Action and Clean Water Fund, California Public Interest Research Group Charitable Trust.

Zakaluk, B.L., and S.J. Samuels, eds. 1988. *Readability: Its Past, Present, and Future.* Newark, Del.: International Reading Association.

ABBREVIATIONS

AWWA	American Water Works Association
AwwaRF	Awwa Research Foundation
CCR	Consumer Confidence Report
CSADW	Campaign for Safe and Affordable Drinking Water
CWS	community water systems
FAQs	frequently asked questions
FOIA	Freedom of Information Act
MCL	maximum contaminant level
MCLG	maximum contaminant level goal
MTBE	methyl tertiary butyl ether
NA	not applicable
NRDC	National Resources Defense Council
NTNCWS	Non-Transient Non-Community Water Systems
OLS	ordinary least squares
PAC	Project Advisory Committee
POU	point-of-use
SDWA	Safe Drinking Water Act
SDWIS	Safe Drinking Water Information System
TNCWS	Transient Non-Community Water Systems
USEPA	U.S. Environmental Protection Agency

IWA Publishing's authorised EU representative for General Product Safety Regulations is Diane D'Arras, 15 rue Duret, 75116 Paris, France, e-mail: safety@iwap.co.uk.

Printed and bound by CPI Group (UK) Ltd, Croydon, CR0 4YY
19/09/2026
02219351-0001